I0463672

The Beautiful
Mechanical Universe

Josef Tsau

ISBN: 978-1-4834-1496-6 (sc)
ISBN: 978-1-4834-1497-3 (e)

Lulu Publishing Services rev. date: 8/8/2014

To the only real science and our families

Both my recently deceased wife Gertrude and I are scientists, and our entire family loves science. Writing this book has been a family affair engaging in many discussions, in particular with my son John. My grandkids Ryan and Anika are happy to offer their picture (front cover) for science. This book is dedicated to the only real science, my family, and yours as well.

Contents

Preface

This is the author's continued efforts to prove to the world that, as natural science has been showing to us for centuries that we live in a beautifully mechanical world where everything is logically explainable, it is the only real science. Its foundation science or mechanical physics is the physics of everything.

He is, however, alone up against the entire physics community that has accepted Einstein's theories of relativity and has led to develop today's mainstream-of-thought physics in the past century. Here, he shows how some old scientific misconceptions have turned to religious beliefs and finally misled the physics community to develop and teach a religious pseudo-physics of four-dimensional universe having its universal phenomena, elementary particles, matters, and so forth discovered by mathematical derivation. Despite mainstream-of-thought physics has been accepted worldwide and considered as the greatest scientific discovery ever by many, the author proves with the strongest scientific evidences possible that it is a fantasy or religious teaching, like heaven or hell not really existing. He shows that it is contradictory to natural science and therefore to experimental findings. Therefore, it cannot and has not been proven correct experimentally, as the physics community claimed. To prove so, the author pointed out that the findings generated by particle accelerators to prove it correct are either invalid or misinterpreted.

Unacceptable through normal channels, self-publishing books has been the author's only way to report his original breakthrough scientific discoveries, which have turned out to be both a broad scope scientific progress and scientific revolution against mainstream-of-thought physics. Under the circumstances, continuous scientific additions and revisions have become necessary to the author's discoveries, forcing him to rewrite his book repeatedly with the latest *Common Sense Physics & Cosmology* in 2010 by Infinity Publishing.com. He is now happy with significant progress made in disproving the entire mainstream-of-thought physics and expanding his new ether theory to understand light, light-related phenomena, and beyond, including disproving the foundation science of the big bang theory.

Besides being the daily applied physics and foundation science of natural science for centuries, the author's breakthrough discoveries in mechanical physics offers logical interpretations of all the scientific topics already interpreted by mainstream-of-thought physics, which therefore have been independently disproved and made obsolete again. The discoveries further prove that mechanical physics is the physics of everything or natural science is the only real science.

The author calls upon the world, the scientific community, the governments of all countries, companies (particularly those relying on science and technologies), and the general public to help end the century-long religious pseudo-science era.

Chapter 1 - Two Physics Problem

Natural science is the method to discover and study the matters and phenomena occurring in nature experimentally. Both Copernicus and Galileo developed it during the sixteenth and seventeenth centuries, leading to the separation of natural science from the religious teachings and beliefs without experimental proof. It has led to the development of mechanical physics, also known as classical physics or Galilean physics, which has long become the foundation science of all fields of natural science such as chemistry, biology, forensic science, and so forth. It started with trial and error, but the scientific community has since discovered many scientific principles to follow, for example, natural science is logical, consistent, and coherent, thus making it important to have accumulative scientific knowledge and to obey discovered scientific laws.

Scientists often encounter the problem of having limited amount of experimental findings, which supports more than one possible scientific interpretation, but only one can be the correct science. Therefore, they often need to design and conduct many experiments to test all scientific possibilities in order to find the only correct scientific interpretation. For example, more than one theory has often been proposed to interpret hard-to-understand scientific issues such as light, gravity, and other universal forces, which have remained to have more than one theory and to be hotly debated scientific issues for centuries.

One long-lasting scientific issue is whether nature has only one type of forces or two. The daily encountered are pushing forces acting on contact, long leading to the development of mechanical physics. In the eighteenth century, Newton introduced his famous universal theory of gravitation proposing that all matters have gravitational force postulated to be attraction force acting from distance. His concepts and postulations not only have since been accepted by the scientific community but have also been used to interpret all other universal forces despite that it is illogical for a force to act from distance and for all matters to have intrinsic gravitational force without identifying the energy source constantly producing it. It has become religion-like beliefs for several centuries while mechanical physics has been developed to interpret everything else and the foundation science of all fields of natural science.

The physics therefore has essentially been split into two since the eighteenth century. The mechanical physics has led to the development of thermodynamics; kinetics; the physics of heat; physical, chemical, nuclear reactions; the discovery of atoms and molecules; and so forth and the foundation science of all fields of natural science.

The belief of universal forces acting from distance remains to be incomprehensible logically. During the nineteenth and twentieth centuries, Faraday introduced force-field concept, offering a way to mathematically interpret universal forces and leading Maxwell to derive the electromagnetic wave equation and to magically find that the theoretical speed of electromagnetic wave matches that of light found experimentally. This magical finding has eventually misled the scientific community to believe that

- light was electromagnetic wave; and
- mathematical derivation could lead to scientific discoveries and welcome Einstein's theories of relativity, upgrading mathematics from a quantification tool to make scientific predictions or discoveries.

The author calls this new physics "mathematical physics" or religious pseudo-physics. In the past century, the scientific community has led Einstein's "scientific revolution" to expand mathematical physics to develop and include the big bang and quantum theories, along with standard model particle physics, to become today's mainstream-of-thought physics.

Although the scientific community has accepted both physics, they are scientifically incompatible with each other; therefore, one must be wrong. This is because that Einstein's theories of relativity are based on several postulations contradictory with the fundamental teachings of mechanical physics or natural science.

For example, mechanical physics teaches that matters have mass, take up space, and have relative motions to collide, interacting with one another. Einstein's theories postulated that light has neither mass nor relative motion. The two physics therefore have their interpretations of everything contradictory with each other.

For example, mechanical physics teaches that matters produce phenomena, meaning that phenomena have physical origins, while mathematical physics teaches that universal phenomena have mathematical origins. And unlike physical origins, they have no mass. Teaching that universal phenomena have mathematical origins should mean that all phenomena have mathematical origins as well. Therefore, wind, fire, and even fingerprints should have mathematical origins, thus having no mass as well.

Because experimental findings discover mechanical physics or natural science, mathematical physics should be contradictory with experimental findings, meaning that it cannot be proven to be scientifically correct experimentally, as the physics community has repeatedly claimed. For example, the mathematical origins of both light and gravity have no mass, and they therefore should be experimentally undetectable and not provable to be light or gravity.

The scientific community has been wrong in teaching that both physics were correct. It teaches that mechanical physics is incapable of interpreting relativistic and quantum phenomena; therefore, both relativity theories and quantum theory have replaced it to interpret the above-mentioned scientific topics. However, because both physics are incompatible with each other, both cannot be correct and therefore cannot coexist.

For example, the author has recently discovered mechanical physics-based scientific interpretations of all the scientific topics already interpreted by mainstream-of-thought physics, but they are all different and contradictory with those interpretations of the mathematical physics. His scientific interpretations, however, do not support the existence of the relativistic phenomena and offer a new interpretation for the quantum effect specific to light spectrum. As a result, the author has exposed the two contradictory physics issues to the open. A brief introduction and discussion of both physics is given below.

Mathematical Physics

In the eighteenth century, Newton's universal theory of gravitation was phenomenally successful in the mathematical calculation of the planets' orbits. As a result, the scientific community has since accepted its postulations that gravitational force is the attraction force of all matters acting from distance. In addition, they have been applied to interpret all other universal forces. Because this type of force is not understandable logically, it has been interpreted only mathematically.

In the nineteenth century, based on the observations that iron powder aligned up around a magnet, Faraday postulated the existence of (matter-free) force fields in space to somehow mediate universal forces acting from distance. Although his concept has not helped logical understanding of universal forces acting from distance, it offered a way for mathematical manipulation.

Based on Faraday's force-field concept, in 1866, Maxwell mathematically derived his famous electromagnetic-wave equation and magically discovered that the calculated propagating speed of the electromagnetic wave matched light speed found experimentally. He therefore proposed that light was electromagnetic wave.

In 1885, Hertz discovered radio wave. Both his conclusions and later findings that lights are produced by electrons have been used as the experimental proofs that lights are electromagnetic wave, leading the scientific community to accept Maxwell's proposal starting an era of religious pseudo-physics believing that mathematical manipulation has discovered the mathematical origin of light and can therefore make scientific discoveries.

A century ago, Einstein introduced his theories of relativity based on both Maxwell's proposal that light was electromagnetic wave and his own postulations that light should have the properties defined by the electromagnetic-wave equation, such as having no mass and an absolute constant speed. His postulations have led to his mathematically derived predictions of the presence of relativistic phenomena. After some of his bold predictions—the strong gravitational force of the sun bending light and the mass of fast-moving particles increasing with mounting speed—had been found supported by some experimental findings, his theories of relativity have shaken the world. They have also gained the acceptance and support of the scientific community and the general public like a religion despite that they were—and still are—far from experimentally proven being scientifically correct.

In the past century, our physics community has been leading Einstein's scientific revolution to develop and teach Einstein's theories of relativity, which have later been combined with quantum theory, standard model particle physics, and the big bang theory to become today's mainstream-of-thought physics. Due to scientific incompatibility between mathematical and mechanical physics, natural science or mechanical physics has been the target of Einstein's scientific revolution. It has long been made unacceptable for use in scientific topics already interpreted by mainstream-of-thought physics, citing its incapability of interpreting relativistic phenomena and quantum effect.

The physics community has been developing and teaching two incompatible physics for centuries, and one must be wrong scientifically. This should have not happened if the scientific principle of the consistency has been obeyed. The physics community should have rejected both Maxwell's and Einstein's propositions that light were electromagnetic wave due to their postulations being contradictory with the fundamental teachings of the natural science that have been repeatedly proven correct.

Also, when more experimental findings become available, the correct science stands tall while the wrong science will be disproved. It has happened over the past fifteen years when the author made breakthrough scientific discovery using mechanical physics to interpret all the scientific topics already interpreted by mainstream-of-thought physics. However, the physics community has repeatedly rejected his discoveries because they have also disproved the entire mainstream-of-thought physics developed and taught by the physics community in the past century.

Undoubtedly besides scientific issues, religious beliefs, prestigious power and positions, and big financial stakes are all playing important roles in creating a religious pseudo-physics era for a century already. It is certainly the biggest scientific mistake ever. The only real science is so important to mankind that we all need to help to bring our scientific community back to science.

Instead of being scientifically proven correct, many concepts—universal forces are attraction forces acting from distance, matter-free force fields exist, and relativistic phenomena exist, and so forth—have been religiously believed to be the proven correct science by the physics community. For example, some reviewers were surprised that the author did not even honor the force-field concept.

Also, the physics community has long been teaching that electromagnetic wave has been "experimentally proven" to be light. It has further been convinced by the proof of the negative finding of Michelson-Morley experiment and the findings obtained by high-energy particle acceleration studies that light has an absolute constant speed and the relativistic phenomena predicted by Einstein's theories of relativity exist. Therefore, it has been teaching that mathematical manipulation discovers universal phenomena.

The above-mentioned concepts are, however, based on postulations contradictory to natural science, and because there have been much more experimental findings to prove natural science, both their postulations and these concepts should have not been accepted to begin with.

The physics community has further expanded the mathematical physics to develop standard model particle physics to mathematically predict the elementary particles making up matters, mediating universal forces, and providing matters with mass (Higgs boson). Again, the physics community has claimed that all predicted elementary particles, except graviton, have been confirmed experimentally.

The mathematical physics therefore teaches that

- mathematical manipulation predicts or discovers both universal phenomena and matters; and
- its predicted or discovered elementary particles have been proven to exist experimentally.

Natural science teaches to study and discover matters; the phenomena matters produce experimentally. Scientists have found that matters are experimentally detectable because they have the unique properties of having mass, relative motions, and the ability to take up space, which makes them colliding, interacting with one another to produce forces and other phenomena experimentally detectable.

However, there is no scientific reason why mathematical derivation can lead to scientific predictions and discoveries. Only God knows what matters and phenomena exist in nature, and if mathematical derivation could discover them, mathematics should be the God. The physics community has therefore turned science to beliefs or religion.

Although quite a few physicists have already claimed the above conclusion, it is apparently unacceptable to the entire scientific community and not even openly suitable to the physics community, which has led Einstein's scientific revolution to develop today's mainstream-of-thought physics making scientific discoveries mathematically.

The physics community has been using experimental methods to prove the discoveries by mathematical physics; therefore, they claimed it is natural science. However, because mathematical physics is contradictory to natural science, it cannot be natural science, and all its interpretations should be contradictory with those of natural science or experimental findings because all the scientific interpretations of natural science are based on experimental findings. Therefore, all the interpretations of mathematical physics should be contradictory with experimental findings due to its contradiction with natural science or experimental findings and the fact is that they cannot be proven correct experimentally. Therefore, the repeated claims of the physics community that mainstream-of-thought physics has been proven correct experimentally are scientifically impossible and invalid.

There have been surprisingly more than expected coincidences misleading our physics community to develop and teach mathematical physics like a religion. Although the entire scientific community appears to support mainstream-of-thought physics, it should not be. Besides mathematical physics, all fields of natural science are mechanical physics-based. Apparently, behind both mysteries and difficult mathematics, mathematical physics is very intimidating; therefore, all other fields of the scientific community have become the silent majority. The scientific community therefore needs to be more confident in that natural science is logically understandable and more vigilant to protect natural science as a whole because any field of natural science is inseparable from other fields.

Mechanical Physics

Mechanical physics is also called "determinant physics" based on collision interactions of matters to produce acting-on-contact pushing forces, energies, and all other phenomena, which are detectable experimentally. Therefore, mechanical physics studies and discovers matters and the phenomena they produce experimentally. It has long been the foundation science of all fields of natural science because they all are based on the scientific knowledge of mechanical physics and study nature experimentally.

Over the past several centuries, steady progress has been made in mechanical physics, including the development of thermodynamics, kinetics, the physics of heat, the discovery of atoms and molecules, the progress of all fields of natural science, and so forth. A century ago, mechanical physics was already considered a well-established fundamental science, except that it still has not been used to interpret such important universal phenomena as light, gravity, and so forth, mainly because the scientific community has its collective mind-set using mathematical physics to interpret them, thus leading to the development of today's mainstream-of-thought physics and a century of Einstein's scientific revolution up against natural science.

The logical conclusion of mechanical physics for the discovery of the presence of gravitational force and all other universal phenomena is that there should be an atmosphere of tiny particles in the universe. For example, consider the "ultra-mundane corpuscles," which Le Sage postulated in the seventeenth century in his gravity theory, capable of penetrating through matters to collide, interacting with all matters to produce gravity (and other universal phenomena). Le Sage's theory, however, has not gained widespread acceptance, and by the twentieth century, it was generally considered discredited apparently due to the scientific community's support of mathematical physics. However, it remained to be the only mechanical physics-based logical scientific interpretation of gravitational force and other universal phenomena.

The physics community has questioned where the tiny particles come from and where they get the energy from to constantly produce gravitational force. It has also argued that, to produce gravitational force, the collision interactions between the postulated tiny particles and matters should be inelastic. Therefore, they should produce large amounts of heat, which, according to the calculation of Maxwell and others, should be so much that it vaporizes matters in seconds. It therefore questions: Where is the heat produced?

The scientific advancement with time has answered the questions of the scientific community on Le Sage's gravity theory. The physics community has been teaching for some time that there is a dense atmosphere of neutrinos constantly produced by all stars from their nuclear reactions, taking up all the space of the universe, including the area taken up by all matters having atomic and molecular structures, because neutrinos penetrate through them like the tiny particles postulated by Le Sage. Besides, the nuclear energy that neutrinos carry can be the energy source to constantly produce gravitational force and other universal phenomena as well. The atmospheric neutrinos, therefore, are likely the tiny particles postulated by Le Sage to collide, interacting with matters to produce gravitational force.

Apparently, the scientific community has essentially ignored this scientific possibility because it would lead to scientific disproving of the entire mainstream-of-thought physics. Also, a major scientific question is still unanswered: If neutrinos collide, interacting with matters to produce gravity (and other universal phenomena), where is the large amount of heat expected by the physics community to be produced by these collision interactions?

Instead, the scientific community teaches and insists that subatomic particles such as electrons and protons are tiny point particles and they are far apart from one another in matters having atomic and molecular structures. Neutrinos freely penetrate through their matters, like vacuum space hardly having any chance to collide, interacting with their subatomic particles. Thus, they have nothing to do with collision, interacting with matters to produce any universal phenomena. This, of course, is the conclusion needed for the scientific community to continue to support its mainstream-of-thought physics.

The Two Contradictory Physics Scientific Community

While continuing to teach mechanical physics, the scientific community has been leading a revolution to develop and teach mainstream-of-thought physics, using it to interpret universal phenomena, cosmology, particle physics, and so forth and to forbid the use of mechanical physics in these same scientific topics apparently due to the incompatibility of the two physics.

In the past century in particular, the scientific community has a corrupted physics community trying to utilize the religious teachings of Maxwell and Einstein to create a kingdom of religious pseudo-physics to control the essentially unlimited scientific R&D and education funding alone with the power and prestige they bring.

It has succeeded in the development of religious pseudo-physics known as mainstream-of-thought physics accepted worldwide like a religion because nobody really understands it. Although it has all the science R&D and education funding worldwide and the power to block the use of mechanical physics, it has difficulties in finding the applications of mainstream-of-thought physics in other scientific fields apparently due to its incompatibility with natural science.

The physics community has been misleading the scientific community as early as the eighteenth century with Newton's theory of universal gravitation followed by Faraday's force-field concept, Maxwell's magical discovery in the nineteenth century that the calculated speed of electromagnetic wave matches that of light found experimentally, and Einstein's theories of relativity in the twentieth century to develop and teach mathematical physics. Mathematical physics is based on beliefs that mathematical derivation leads to scientific discoveries. It is not natural science because it is incompatible with natural science due to both that its fundamental postulations contradict the fundamental teachings of natural science and the difference in their method to make scientific discoveries.

However, mainstream-of-thought physics has been wrongly regarded as natural science worldwide by the repeated claims of the physics community that the predictions or discoveries of mathematical physics have been proven correct by experimental findings. Because it is contradictory with natural science, it is scientifically impossible for mathematical physics to be proven correct experimentally. In this book, it will show that the experimental findings used to prove mathematical physics by the physics community are either invalid or misinterpreted.

The following sketch shows that, for several centuries, our physics community has been corrupted to develop and teach two physics contradictory with each other.

Two physics problem

Natural science: Discover the laws of nature experimentally.

During 16-17 Centuries both Copernicus and Galileo have discovered natural science, led a scientific revolution to separate science from religion, and led the development mechanical physics.

In 18 Century Newton has split physics into two!

Newton's three laws of motion

Mechanical physics: Pushing forces acting on contact, energies, and other phenomena are produced by collision interactions of matters.

The development of thermodynamics, kinetics, physical, chemical, nuclear reactions, atoms, molecules, and other fields of natural science.

Tsau's breakthrough discovery: mechanical physics based Interpretation of universal phenomena, particle physics, cosmology, etc.

Mechanical physics is the physics of everything.

Newton's universal theory of gravitation postulates that gravitational force is the attraction force of matters acting from distance.

Mathematical physics: It mathematically interprets universal forces postulated to act from distance.

Mathematical physics has been expanded to include Faraday's force-field concept, Maxwell's electromagnetic-wave theory, Einstein's theories of relativity, big bang theory, quantum theory, and standard model particle physics to be today's mainstream-of-thought physics.

A religion believes and teaches that mathematical derivation discovers both universal phenomena and elementary particles making up matters and mediating universal forces.

Chapter 2 - **Mathematical Physics**

In the eighteenth century, Newton introduced his universal theory of gravitation postulating that that all matters possess gravitational force, an attraction force capable of acting from distance to interact with one another. Due to the phenomenal success of this theory in the calculation of orbits of planets, the scientific community has accepted it, including its postulations. Ever since, the scientific community has also postulated that all universal forces were forces acting from distance, which are not logically understandable but mathematically interpretable. Although Newton's universal theory of gravitation is illogical, its mathematics have been proven correct experimentally for the interpretation of the planets' orbits. However, it cannot specify what the real science is. For example, it supports Newton's, Einstein's, and Le Sage's postulated science as well. Among the three proposed sciences, only Le Sage's gravity theory is logically understandable. It postulates that gravitational force is an acting-on-contact pushing force produced by atmospheric tiny particles due to their collision interactions with all matters.

Unfortunately, Newton's postulations that universal forces are forces acting from distance has become a religious belief, leading to Faraday's force fields concept and Maxwell's accidental and magical discovery that the propagating speed of electromagnetic wave matches that of light found experimentally to further lead to Einstein's theories of relativity teaching, among others, that mathematical derivation leads to scientific discoveries. It has been expanded to include modern cosmology and standard model particle physics to become today's mainstream-of-thought physics. The author also calls it "mathematical physics."

Einstein's Theories of Relativity

In the nineteenth century, Maxwell suggested that the electromagnetic wave equation, mathematically derived based on Faraday's postulations that matter-free force fields existed in space to somehow mediate universal forces acting from distance, represented light because its calculated speed magically matched that of light found experimentally. In the early twentieth century, Einstein published his theories of relativity, which further postulated that light should have the properties defined by the electromagnetic-wave equation. Both Maxwell and Einstein therefore started a new physics based on believing that mathematical equations can define or represent universal phenomena, meaning that universal phenomena have mathematical origins discoverable by mathematical derivation. The author calls it "mathematical physics" in this book.

To meet the requirements of the electromagnetic-wave equation, Einstein postulated that light has no mass traveling at an absolute speed, leading to his mathematical conclusions or predictions that relativistic phenomena exist. Accordingly, both time and dimensions of matters are not absolute but a function of the content and moving speeds of environmental matters and that the mass of matters increases with their moving speed to infinitely large upon reaching light speed. The physics community claims that some of the relativistic phenomena, such as that the mass of charged particles increase with their moving speed, have already been confirmed experimentally in high-energy particle acceleration experiments.

Einstein's theories of relativity, however, contradict with the fundamental teachings of mechanical physics, such as that all phenomena have physical origins. For example, the phenomena we encounter daily—wind, fires, fingerprints, diseases, and so forth—are all known to have physical origins, and none has mass-free mathematical origin.

The negative finding of the Michelson-Morley experiment, over a century old, has been used to prove Einstein's postulations that light has an absolute constant speed and there is no ether in space to transmit light. It has also been convinced by the findings obtained in high-energy accelerators, which can be interpreted by that the mass of charged particles increases with their increasing moving speed and they cannot be accelerated to reach light speed, all as Einstein has predicted.

Einstein's general theory of relativity mathematically interprets the universe with (force) field equations. It is a four-dimensional universe. Besides three dimensions, time is also a variable. According to Einstein, the old space is now spacetime, which environmental matters and their moving speeds can somehow bend. He teaches that curved spacetime somehow produces gravity.

The Big Bang Theory and Modern Cosmology

Modern cosmology is based on both the general theory of relativity and the big bang theory. The big bang theory is based on the scientific interpretation of the phenomenon of the universal redshift of starlight by that stars are leaving one another and therefore the universe is uniformly expanding. It suggests that, as we go back in time, the universe gets smaller, denser, and hotter. The big bang theory therefore postulated that the universe was born from a big explosion or bang of something infinitely small, dense, and hot called "singularity" about fifteen billion years ago, which has been uniformly expanding ever since.

The general theory of relativity teaches that gravity is curved spacetime and gravitational force is proportional to the mass content of matters and can be infinitely strong. It further teaches that there is no ether or nothing in space; therefore, gravitational force controls the motions of all stars and their planets in galaxies and among galaxies. The future of the universe therefore depends on the total mass density of the universe to expand forever, to gradually stop expanding, or even to contract to finally collapse.

Without having ether or anything else in space to collide, interacting with light, the big bang theory is the only possible scientific interpretation of both the universal redshift of starlight and the Hubble Law. It has predicted that there should still be residue radiation left from the big bang; therefore, the scientific community was happy to discover cosmic microwave interference background (CMB), suggesting it is the residue radiation from the big bang, confirming the theory's prediction.

However, it has been well established knowledge that matters emit lights and heat and therefore CMB should also be emitted by matters instead of being the residue radiation of the big bang. Because CMB comes from all directions, it is likely radiated by matters inside the intergalactic space surrounding our galaxy, where, without having a star, it has a very low and uniform temperature, approximately 2.7 degrees Kelvin.

It has been known for over a half century that there is an atmosphere of neutrinos constantly produced by the nuclear reactions in stars. The scientific community teaches that matters having atomic and molecular structures have point-sized subatomic particles far apart from one another and neutrinos hardly have any chance to collide, interacting with matters and therefore penetrating right through them. Therefore, neutrinos are not ether and do not collide, interacting with matters to produce gravitational force or any other universal phenomenon. The planets of the solar system appear to have constant orbits around the sun, which offers the strongest scientific evidence to prove that the atmosphere of neutrinos has no drag on moving matters.

The theory further teaches that gravitational force is the only universal force to shape up the universe. It collects cosmic hydrogen and helium produced by the big bang to form the first-generation stars, and when they are big enough, their strong gravitational force starts nuclear reactions at their centers. When nuclear fuels, hydrogen and helium, are used up, the strong gravitational force will crash stars into (dead) dense stars such as neutron stars and white dwarfs. It further teaches that all stars, including dense stars, have strong gravity and the gravity of black holes are so strong that even light cannot escape from them.

Astronomers have found many spiral galaxies having their stars orbiting around their centers. If it is the gravitational force alone holding them together without flying apart, according to calculations, these spiral galaxies need a lot more matters than the visible stars and their planets to produce such strong gravitational force. The scientific community therefore teaches that the universe contains mainly undetectable or dark matters to produce the gravitational force strong enough to keep spiral galaxies intact and the cosmos together.

To explain why the universe is constantly expanding, physicists further postulated the presence of undetectable dark energy to pull the universe apart. Based on their theoretical calculations, there should be only 5 percent visible or detectable matter in the universe, 27 percent dark matters, and 68 percent dark energy. Therefore, in total, 95 percent are the dark matters and dark energy, undetectable and invisible.

Quantum Theory

The findings that the light spectrum (energy) of a black body was not continuous but discrete lines of energies having the so-called quantum levels led Max Planck to develop quantum theory in 1900. He postulated that energies were quantized in a microenvironment. Because orbital electrons of atoms and molecules emit light, quantum theory suggested that their orbital are quantized, which has led to mathematical interpretation of atomic and molecular structures. Yet, quantum effect may be a unique property of light not necessarily applicable to all other energies in the microenvironment such as universal forces, which apparently show no quantum levels. However, because light has been postulated to be electromagnetic wave, the electromagnetic force field should be quantized to produce quantized light spectrum. And as a result, all universal forces should also have quantum effect, thus leading Dirac and other physicists to develop quantum electrodynamics (QED) theory to wed the classical physics-based electrodynamics with quantum theory and even with the theory of relativity. Consequently, light has also been called photon, which is both wave and particle, and has been assigned to be the particle mediating electromagnetic force. Therefore, quantum theory teaches that all energies are quantized in the microenvironment. The theory also started the principle of the force-particle duality: every (universal) force has its special force (mediating) particle.

Unable to interpret both relativistic and quantum effects mechanical physics has long been disallowed for use in such scientific topics as universal phenomena, cosmology, particle physics, and so forth. However, the scientific community continues to teach mechanical physics, and it continues to be the daily applied physics. Furthermore, the progress of all other fields of natural science, which are all mechanical physics-based, are unstoppable despite that Einstein's scientific revolution has made mechanical physics unacceptable in many important scientific topics.

Standard Model Particle Physics

It was formerly called "elementary particle physics." To begin with, there were only three known elementary particles: electron, proton, and neutron. To wed general relativity with quantum theory, Dirac developed relativistic quantum mechanics, and his theory predicted anti-electron or positron, which was found two years later in 1933 in both cosmic ray study and high-energy particle collision studies. The concepts of antiparticle and antimatter, as all *Star Trek* fans know, is that, when a particle and an antiparticle collide, they annihilate, converting 100 percent of their mass into energy, obeying Einstein's equation. In 1937, muom was discovered, and in 1947, pion and lambda were found. In 1953, a powerful particle accelerator was built, and suddenly, the avalanche of the discovery of elementary particles was unleashed. By the 1960s, about a thousand of them have been discovered.

Ever since, the physics community has changed the name of elementary particle physics to "particle physics." In 1963, Murray Gellman proposed his theory of quantum chromodynamics, suggesting the presence of (six) quark particles and their boson (gluon) to make up all known baryons (heavy particles such as neutrons and protons) and mesons (medium heavy particles such as pion), which has led to the great simplification of particle physics to shape today's standard model particle physics, which, of course, are also based on force field concept, theories of relativity, and quantum theory. Together with many postulations of its own, it has to be the weirdest part of mainstream-of-thought physics.

Standard model particle physics is the theoretical framework describing all the currently known elementary particles, including Higgs boson. It postulated the presence of six flavors of quarks, named up, down, strange, charm, bottom, and top, along with their antiquarks to make all elementary particles. There are elementary matter particles and antiparticles or fermions (quarks, leptons, antiquarks, and antileptons) and force particles or fundamental bosons (gauge bosons and Higgs boson) that mediate interactions among fermions.

Force particles are forces between particles that arise from the exchange of other particles. These force carrier particles are bundles of energy (quanta) of a particular kind of field. There is one kind of field for every species of elementary particle. For example, there is an electron field whose quanta are electrons and an electromagnetic field whose quanta are photons. The force carrier particles that mediate the electromagnetic, weak, and strong interactions are gauge bosons.

Wave-particle duality is standard model particle physics describing nature in terms of fields. Each field has a complementary description as the set of particles of a particular type. A force between two particles can be described either as the action of a force field generated by one particle on the other or in terms of the exchange of virtual force carrier particles between them.

The energy of a wave in a field (for example, electromagnetic waves in the electromagnetic field) is quantized, and the quantum excitations of the field can be interpreted as particles. The standard model contains the following particles, each of which is an excitation of a particular field:

- Gluons, excitations of the strong gauge field
- Photons, W bosons, and Z bosons, excitations of the electroweak gauge field
- Higgs boson, which gives mass to elementary particles

Gravity is not considered a part of the standard model, but it is thought gravitons may exist as the excitations of gravitational waves. The status of this particle is still tentative because the theory is incomplete and the interactions of single gravitons may be too weak to be detected.

The scientific community claims that its standard model particle physics has predicted many elementary particles and then confirmed its predicted discoveries experimentally. Among them are positron and other antiparticles, the force particles of universal forces such as pion mediating strong nuclear force and W mediating weak nuclear force, quirks, and gluons. The most famous predicted particle of all of course is the Higgs boson, nicknamed "God particle," offering all particles and matters their masses.

On the birthday of Unified States in 2012, physicists in CERN announced the discovery of a particle presumed to be the Higgs boson, which has been hailed to be the biggest physics news of the twenty-first century. Therefore, besides the derivation of the mathematical origins of universal phenomena, the scientific community further claims that mathematical physics has predicted the elementary particles both making up matters and producing universal forces. The scientific community therefore essentially teaches a religion that mathematics is God, capable of predicting (creating) both (universal) phenomena and matters. Because the scientific community claims that mathematical physics is natural science because its discoveries have been proven experimentally, this means it has proven natural science to be a religion.

Chapter 3 - **Disproving Mathematical Physics**

Not Natural Science

Both Copernicus and Galileo have discovered the natural science that discovers the laws of nature experimentally. What experiments do is detect and study matters in different ways, such as detect the forces or energies they produce and observe the light they emit or reflect. Matters are experimentally detectable because they have mass and relative motions and take up space, and with all these unique features, they collide, interacting with one another to produce phenomena, which are detectable experimentally and the foundation science of natural science.

Natural science therefore studies matters. Anything without mass, relative motions, or the ability to take up space cannot be experimentally detectable and thus is not natural science. Although very small particles such as molecules, atoms, neutrinos, and so forth may be undetectable individually, the phenomena they collectively produce are often detectable. Also, undetectable particles and their phenomena may become detectable with an improved detection method. However, once experimentally detectable, they have to be matters or the phenomenon they collectively produced.

The teachings of mainstream-of-thought physics contradict with natural science such as that mathematical origins have no mass, light has an absolute speed without relative motions, and it makes scientific discovery by mathematical derivation not experimentally. It therefore is not natural science as generally believed.

Not Provable Experimentally

Both the scientific community and worldwide governments have been developing and teaching Einstein's theories of relativity-based mainstream-of-thought physics in the past century in the name of natural science. The scientific community has convinced the world by repeated claims that its discoveries have been proven correct experimentally.

However, it is scientifically impossible to prove mainstream-of-thought physics correct experimentally because it is contradictory with natural science and therefore with experimental findings as well. It means that the scientific interpretation of everything by the two physics should be different and contradictory with each other. Because natural science is proven experimentally, all the interpretations of mainstream-of-thought physics should be contradictory with experimental findings and therefore cannot be proven experimentally.

For example, mainstream-of-thought physics teaches that (universal) phenomena have mathematical origins, while mechanical physics teaches that phenomena have physical origins having both mass and relative motions. Without having mass or relative motions to collide, interacting with one another to produce phenomena, the mathematical origins should not be experimentally detectable and therefore cannot be studied or proven experimentally. If universal phenomena have mathematical origins, all phenomena should also have mathematical origins; therefore, the interpretation of all phenomena by mainstream-of-thought physics is different and contradictory with those of mechanical physics.

Because light is experimentally detectable and cannot be electromagnetic wave, there is no more scientific reason to postulate that relativistic phenomena exist and to believe that mathematical derivation can predict or discover both universal phenomena and elementary particles. Therefore, none of the mathematically-derived relativistic phenomena, mathematical origins of universal phenomena, and elementary particles should really exist and be proven to exist experimentally as the physics community has claimed.

According to the claims of the scientific community, many scientific predictions or discoveries made by mathematical derivation have been proven experimentally. Because it is scientifically impossible to do so, all the experimental findings used to prove mainstream-of-thought physics must either be invalid or misinterpreted.

What we are talking about are century-long accumulated R&D findings worldwide, and it is highly unlikely that they all can be wrong or misinterpreted. However, it may be possible that the scientific community has been relying on one type of experimental method too heavily and the method somehow repeatedly produces invalid findings. This has been found to be the case, and the particle accelerator has been identified to be the experimental method by the author. He also points out that the accumulative scientific misconceptions, such as the beliefs that universal forces are attraction forces acting from distance, have played an important role in misinterpretation of experimental findings as well. The proofs of invalid and misinterpreted findings generated by particle accelerators are given in many scientific topics throughout the book.

Nonexistence of Relativistic Phenomena

Since the eighteenth century, the scientific community has postulated that all universal forces are forces acting from distance. In the nineteenth century, Faraday postulated that mass-free force fields exist in space to somehow mediate universal forces acting from distance. Based on the postulation of the existence of both electric and magnetic force fields, Maxwell mathematically derived the electromagnetic-wave equation and magically found that the calculated speed of the electromagnetic wave matched that of light found experimentally. He therefore proposed that light was electromagnetic wave.

In the early twentieth century, Einstein further postulated that light has no mass and has an absolute speed to fit light's properties to the electromagnetic wave equation, leading to the development of his theories of relativity, which predicted the presence of the relativistic phenomena from the mathematical conclusions of his postulations.

The theories teach that both time and dimensions are not absolute but changing with both the environmental matters and their moving speeds. Also, the mass of a matter is not constant but increases with moving speed of the matter to infinitely large at light speed; therefore, nothing can reach light speed. Also, fast-moving matters slow down time, and the higher the speed, the slower the time, which stops at light speed. Upon approaching a massive heavenly body, time also slows down, thus leading to producing gravity also known as "curved spacetime." Making time a variable of environmental matters and their moving speed has changed the three-dimensional "classical" universe to a four-dimensional spacetime universe. All these are the so-called relativistic phenomena.

Because light is experimentally detectable, it cannot be electromagnetic wave, which has no mass and no relative motion and therefore should not be detectable experimentally. Therefore, light cannot be electromagnetic wave, and there is no more scientific reason to postulate that light is electromagnetic wave, has no mass, and has an absolute speed particularly when these postulations are scientifically incompatible with the fundamental teachings of natural science that all matters have mass and relative motions. Therefore, there is no scientific reason for all relativistic phenomena to exist.

However, the scientific community has been very sure that some relativistic phenomena have been experimentally proved to exist. For example, it believes that R&D findings obtained in high-energy particle acceleration studies have proved that the mass of charge particles increases with increasing their moving speed and these particles cannot be accelerated to light speed, just as Einstein had predicted. However, the scientific community still does not know what the charge energy both electron and proton constantly carry is and whether it is stable to motions and accelerations.

Because light cannot be electromagnetic wave, the matching of the theoretically calculated speed of electromagnetic wave with light found experimentally should be only a coincidence. Besides, Hertz's conclusions and later findings that electrons, such as the orbital electrons of atoms and molecules, produce light; however, these have not proved that light is electromagnetic wave or anything else. Because light is an experimentally detectable phenomenon, it should be produced by particle having mass, not wave without mass, which does not have any scientific meaning. Light cannot and has not been proven to be electromagnetic wave; therefore, relativistic phenomena have no scientific reason to exist. Some of Einstein's bold predictions, such as that strong gravitational force can bend light, have been timely proved by limited amount of experimental findings, which, however, have not scientifically proved his theories of relativity.

Both spacetime and curved spacetime are part of relativistic phenomena as well. Einstein postulated that curved spacetime produces gravity, but he has not offered a scientific understanding of gravity. His prediction that strong gravity bends light is therefore just a lucky guess. Again, without having mass, spacetime and curved spacetime are not natural science. And because gravity is detectable, it cannot be curved spacetime, which having no mass should not be detectable experimentally. Again, because light cannot be electromagnetic wave, both relativistic phenomena and spacetime should not exist.

The effort to prove that time is not absolute so far has been inconclusive and controversial. Besides, there is no reliable method to measure absolute time, not even atomic clocks. Mainstream-of-thought physics teaches that atomic clocks may go a lot faster in collapsing stars due to their increased instability, proving the rates of nuclear decay change with its environment. Therefore, there is no reliable experimental method to measure absolute time. Mechanical physics also teaches that the stability of atomic nuclei change with environment such as temperature.

It is true, as the scientific community claims, that mechanical physics cannot interpret relativistic phenomena, but it is because of their nonexistence. The breakthrough discoveries of the author given in this book show that mechanical physics does not need relativistic phenomena and their postulations to interpret universal phenomena, cosmology, particle physics, and all other phenomena, making the entire mainstream-of-thought physics obsolete.

Disproving Standard Model Particle Physics

Like the entire mainstream-of-thought physics, standard model particle physics is based on centuries-old postulations that universal forces act from distance and Faraday's force-field concept, which have long become religious beliefs. Later, force-particle duality principle has been introduced to somehow mediate universal forces acting from distance, postulating or predicting that all universal forces have their own mediating particles. The scientific community has claimed that the theories of the standard model particle physics have predicted all the elementary particles making up matters, mediating universal forces, and producing mass. They also say that all except the graviton, predicted to mediate gravitational force, have been confirmed experimentally.

Recently, the scientific community has announced its

experimental confirmation of a long-predicted mass-producing particle known to be Higgs boson, or the "God particle." Its experimental confirmation has been by only one type of experiment, particle accelerator, based on a very weak signal, which may indicate something there, but it has not proved that it is Higgs boson or anything else. In fact, particle accelerators have been used to confirm if not all most of the predicted elementary particles predicted by standard model particle physics, similar to the confirmation of Higgs boson.

If for some reason the particle-accelerator method is found wrong or invalid, the experimental confirmation of all predicted elementary particles will be invalidated, and this happens to be the case, as it will be given in the following section. Like Higgs boson, the force particles of universal forces predicted by standard model particle physics have also been found in particle accelerators at the mass range predicted mathematically. Therefore, due to invalid experimental proofs, these elementary particles have not been confirmed to exist. Besides, none of them have been found to be present in space as free and stable particles, like air or neutrino particles readily available for producing (or mediating) universal forces anywhere, which have been postulated to act from distance forces. Without being present in space, how can force particles mediate universal forces?

Positron has been one of the theoretically predicted elementary particles, and its existence has been confirmed experimentally. It has been found in the residue of cosmic ray and in the emissions of radioactive nuclei. Physicists can also make some light atoms radioactive to emit positrons and have found some important medical applications with them. As its name indicates, it was predicted to be an antiparticle, and its discovery has led to the claim that both antiparticles and antimatters exist.

Although it has been known that matter and antimatter collide to annihilate into 100 percent energy, the positrons emitted from nuclei should hit orbital electrons, but they don't seem to produce huge amounts of heat as expected.

The author has a new scientific interpretation for positron.

He has found that the content of both neutrinos and ethers of an electron varies with the environment it is in. For example, free static electrons contain the most, while the electrons inside atomic nuclei contain the least. Therefore, the electrons inside nuclei are highly deficient in the content of neutrinos and ethers resulting in having the properties of "positron." They have been found unstable because, upon emitting from atomic nuclei, they quickly capture environmental neutrinos and ethers to turn back to normal electrons. The existence of both antimatters and their antiparticles therefore has not been proven experimentally as claimed.

The standard model particle physics postulates and teaches that both proton and neutron in atoms and molecules are made of quarks and gluons. Again, there is no creditable experimental finding to support this theory and to prove that both quarks and gluons really exist.

The author's breakthrough discovery shows that both the atmospheric neutrinos and ethers constantly produced by all stars are the common physical origin of all universal phenomena. For example, he interprets that the charge energy of both electron and proton is produced by a dense, turbulent dust cloud of neutrinos and ethers surrounding them. Still, the particle physics of mechanical physics only has four elementary particles (the neutral particles inside both proton and electron, the neutrinos, and the ethers) to make up all matters and to produce all universal forces, which are pushing forces acting on contact. Accordingly, both electron and proton carry charge energy because they have structural features and cannot be further divided to still carry charge energy.

Also, none of the elementary particles predicted by standard model particle physics can find its place in the particle physics of mechanical physics. For example, mechanical physics teaches that all universal forces are produced by both neutrinos and ethers and they are pushing forces acting on contact. The physics of attraction forces acting from distance-based mainstream-of-thought physics therefore has been found both obsolete and scientifically disproved.

Fraudulent Experimental Method

Because mainstream-of-thought physics is contradictory with natural science, it cannot be and is impossible to be proven experimentally, as the physics community claims. Therefore, it is reasonable to suspect that there should be a major scientific problem in the experimental methods used to confirm its predicted elementary particles. Because the physics community has long been heavily relying on particle accelerators to experimentally confirm and develop mainstream-of-thought physics, this type of experimental methods likely has the suspect scientific problems.

Although particle accelerators have discovered over a thousand elementary particles, the science that all matters are made of only two elementary particles, the proton and electron, still stands. It means that both the physics and the chemistry developed in the past, such as the interactive properties between protons and electrons to form atoms, molecules, and so forth, are still the correct science. Findings show that both electron and proton are unique among all charge particles discovered. For example, they are the only two stable charge particles among about a thousand found, and they are also the only two that collide, interacting with each other to form atoms and molecules. It should be expected that the unique size and/or mass ratio of the two affect the interaction properties of the charge energies between an electron and a proton.

For example, proton attracts electron but repulses proton, and an electron repulses other electron. Therefore, the interaction properties between the charge energy of a proton or electron and the more than thousand charged particles found by particle accelerators are unknown and cannot be the same as those found between and among electrons and protons.

A scientific instrument must be calibrated and validated before the scientific community can accept it. For example, a mass spectrometer has been calibrated by ions having known masses and charge energies for the quantification of the mass contents and charge energies of unknown ions. Particle accelerators should also be properly calibrated before use.

As discussed previously, the collision interaction properties between protons and electrons are very unique or specific, which should be very different from their collision interaction properties with the other thousand charge particles. Therefore, we cannot use instruments made of electrons and protons to quantify the charge energies and mass contents of the elementary charge particles other than proton, electron, and their atomic and molecular ions unless we can properly calibrate the particle accelerator first.

However, none of the mass and charge energy contents of the elementary particles predicted by the theories of the standard model particle physics is known. Therefore, there is no way to properly calibrate particle accelerators to quantify or even to reasonably estimate the mass and charge-energy contents of the elementary particles other than proton and electron.

As a result, the experimental findings generated by particle accelerators the physics community used to prove mainstream-of-thought physics are invalid because the particle accelerators used have not been properly calibrated for the intended R&D uses. Therefore, none of the predicted elementary particles by the standard model particle physics has been proven to exist experimentally, as the physics community claims.

Proven Scientifically Wrong and Obsolete

As expected from the incompatibility between mechanical physics and mainstream-of-thought physics, all the mechanical physics-based interpretations are both different from and contradictory to those of mainstream-of-thought physics. For example, all the phenomena of mechanical physics have physics origins, while all phenomena of mainstream-of-thought physics should have mathematical origins, not only universal phenomena. If mainstream-of-thought physics were correct, the mechanical physics-based interpretations of universal phenomena, cosmology, particle physics, and so forth, which mainstream-of-thought physics have already interpreted, should never be found. Yet, the author's breakthrough scientific discoveries offer exactly all these interpretations given in chapters five and six.

Mechanical physics has been repeatedly proven correct for centuries, along with the foundation science of all fields of natural science, and now with the author's breakthrough discoveries, it is proven to be the physics of everything or that natural science is the only correct or real science. It further means that mainstream-of-thought physics is proven scientifically wrong or obsolete.

Disproving the Big Bang Theory and Its Cosmology

The big bang theory teaches that the universe is uniformly expanding, and it is based on the scientific interpretation of the phenomenon of universal redshift of starlight and Hubble Law. However, this same interpretation cannot be applied to interpret other similar phenomena of light such as gravitational redshift and bending of light. Besides, it is not the only possible scientific interpretation, as the scientific community claims, because the presence of all the above-mentioned phenomena of light also suggests that there may be an atmosphere of tiny particles like ether in space to collide, interacting with light to produce these phenomena. The physics community has long claimed that the Michelson-Morley experiment has disproved the presence of ether in space. However, the conclusions of this experiment have been invalided by that light is not (electromagnetic) wave and light should be tiny particles, which hardly collide or interfere with one another in space to change the so-called light-interference pattern.

The other possible explanation of the universal redshift of starlight is the author's "new ether theory," suggesting there are particles even much tinier than light and neutrinos in space, which should be simultaneously produced with neutrinos by the nuclear reactions in all stars to collide, interacting with light particles in space to produce all the discovered interactive universal phenomena of light, and to explain many other universal phenomena. Because it can also consistently and coherently explain gravitational redshift of light, gravitational bending of light, and many other universal phenomena besides universal redshift of light, it should be the scientifically correct theory instead of the big bang theory to explain it alone. The presence of an atmosphere of ether particles also leads to explain what light is, why the light spectrum shows quantum effect, and even why both proton and electron have a dense, turbulent dust cloud of neutrinos and ethers to produce charge energy and to form atoms, molecules, and so forth given in chapters five and six.

The big bang theory is the foundation science of modern cosmology. Apparently, it will be very hard for the scientific community to face the fact that it is scientifically wrong. The author, however, has also discovered a new cosmology consistent with mechanical physics also given in chapter six, offering another strong scientific evidence proving the big bang theory scientifically wrong and obsolete.

Cosmic Microwave Interference Background

Many physicists have been excited by the discovery of cosmic microwave interference background (CMB) because the big bang theory has predicted that the big bang giving birth to the universe should still have some residue radiation left by now. Therefore, they welcome the discovery of CMB as another successful prediction of mainstream-of-thought physics.

It is well-established science that matters emit lights, including heat. Therefore, there is no scientific reason to suggest why CMB is not radiated by matters like all other lights. Like a black body radiates a typical light spectrum characteristic of its own particular temperature range, CMB should be emitted by matters in an environment having very low and uniform temperature, approximately 2.7 degrees Kelvin. Matters inside our galaxy cannot emit CMB because it has many stars distributed throughout the galaxy, making its temperature unevenly distributed. Because CMB comes from all directions, it is likely emitted by the matters inside the intergalactic space surrounding our galaxy, where, without having any star, it should have a uniformly low temperature.

An intergalactic space may have many dead stars, planets, and asteroids, and their CMB radiations should be "grainy," but CMB signals are smooth. In the extremely low temperature environment, the crystalline structures of solids have been shattered. Besides, because there are many winds of neutrinos and ethers coming from surrounding galaxies, many collisions may occur inside an intergalactic space, which may have broken up the heavenly bodies inside it. As a result, the matters inside an intergalactic space may have most broken up heavenly bodies, thus, unusually strong CMB radiation, not as grainy as expected. Also, an experimental method used to detect the weak CMB energy may also contribute to the smoothness of the CMB signal.

NASA has discovered that CMB shows anisotropy property: the CMB coming from southwest having stronger energy than CMB coming from northeast. This is unexpected by the physics community because CMB coming from all directions happens to be nature's design of the Michelson-Morley experiment. Because the it has proved that light has an absolute speed, CMB should not have shown anisotropy property. Nature, therefore, has proven that the detecting method of the Michelson-Morley experiment is invalid or light is not wave.

NASA's findings may suggest that

- the entire galaxy is moving inside the intergalactic space surrounding it to produce the anisotropy observed; *or*
- the anisotropy property of the CMB shows that the solar system is located far from the center of Milky Way galaxy and it is significantly closer to the intergalactic space of the southwest side of the galaxy than that in the northeast side.

Disproving the Physics of Forces Acting from Distance

Both Einstein's theories of relativity and the physics of (attraction) forces acting from distance are the foundation sciences of mainstream-of-thought physics. The author's breakthrough scientific discoveries show that atmospheric neutrinos and ethers are the common physical origin of universal phenomena and all universal forces are the pushing forces of neutrinos and ethers acting on contact. The findings have scientifically disproved the presence of universal forces acting from distance. Therefore, there is no need of Faraday's force-field concept, which led Maxwell to derive the electromagnetic wave equation and further led to the development of Einstein's theories of relativity and mainstream-of-thought physics. The proof that mechanical physics is the physics of everything or natural science is the only real science disproves the entire mathematical physics.

The author's breakthrough discovery should be used to replace the entire mainstream-of-thought physics. Besides, it offers a new level comprehensive scientific understanding of the nature and brings exciting new challenges to the scientific community, such as the possible existence of mini-atoms and mini-compounds.

Chapter 4 - **Major Scientific Issues**

The Collision Interactions of Neutrinos with Matters

The major scientific question is whether it is a mechanical universe that makes discovery experimentally or a mathematical universe that makes discovery by mathematical derivation. The physics community has been teaching that both experimental and mathematical derivation methods make scientific discoveries because it claims that experimental findings have validated many mathematically derived discoveries. However, the fact is that mathematical physics is developed with postulations contradictory with the fundamental teachings of mechanical physics or natural science. Therefore, it is either a mechanical universe or a mathematical universe, not both.

Experimental method can detect matters and their phenomena because matters have mass and relative motions and take up space. They undergo collision interactions to produce forces, energies, and other phenomena detectable experimentally. Mechanical physics teaches collision interactions of material particles, which are detectable experimentally and/or observable visually, produce acting-on-contact pushing forces and all phenomena. It is also known to be the determinant physics that can logically find out what matters produce what phenomena and explain why and how they happen. The study and discovery of matters and their phenomena experimentally have long been called natural science.

There, however, is no scientific reason to suggest that mathematical derivation is able to make scientific discovery of matters and their phenomena. Maxwell magically discovered that the speed of the electromagnetic wave matches that of light; therefore, he proposed that light is electromagnetic wave. However, light and all other universal phenomena are experimentally detectable; therefore, they have been proven to be produced by particles or matters. So they cannot be electromagnetic wave, curved spacetime, and so forth, which have no mass and should not be experimentally detectable.

Einstein's theories of relativity are based on postulations that light were electromagnetic wave having the properties defined by the electromagnetic-wave equation and therefore light has no mass traveling at an absolute speed. Having no mass means nothing or nothing can be detectable experimentally. If light were electromagnetic wave, it should be undetectable. The scientific community has long been teaching both that universal phenomena such as light and gravity are nothing (having no mass) but are detectable experimentally.

Mathematical physics teaches that universal forces acting from distance, which led Faraday to develop the force-field concept and Maxwell to derive the electromagnetic wave equation magically discovering that the calculated propagating speed of electromagnetic wave matched that of light found experimentally. This finding eventually led the scientific community to the acceptance that light was electromagnetic wave, Einstein's theories of relativity, the expansion of mathematical physics to today's mainstream-of-thought physics, and Einstein's scientific revolution against mechanical physics in the past century.

Meanwhile, mechanical physics has been expanded or branched into many fields of natural science. Besides some scientific topics of few physics and related scientific fields such as physics and cosmology have been interpreted by mathematical physics, the majority of the fields of natural science are still essentially unaffected by Einstein's scientific revolution apparently due to the intrinsic controversy or inapplicability of the mathematical physics to them. For example, mathematical physics finds no application in forensic science, which seeks physical evidences to protect human rights. The inapplicability of the mathematical physics to other scientific fields is a good indication that

- mainstream-of-thought physics is incompatible with natural science and should be scientifically wrong; and
- its interpretation of universal phenomena and cosmology should also be wrong.

By the end of the nineteenth century, mechanical physics had long been repeatedly proven to be the correct physics and the foundation science of all fields of natural science. Therefore, it had been a major scientific mistake for the physics community to support Einstein's theories of relativity based on postulations contradictory with mechanical physics or natural science.

In the twentieth century, those in the physics community supporting Einstein's theory and his scientific revolution had quickly taken control and run the physics community like a religion because Einstein's theory teaches beliefs without scientific understanding. For example, it teaches that mathematical derivation makes scientific discoveries and experimental findings have proven these discoveries but does not explain why it can do so.

Mathematical derivation discovered electromagnetic wave, which according to the physics community, has been proven to be light. Also, Einstein's derived field equations to describe the universe, gravity, and standard model particle physics has discovered many elementary particles making up matters, mediating universal forces, and even making mass. They all have been proven correct experimentally. It is religious teaching because nobody knows why mathematical derivation makes scientific discoveries. However, it should also say science because it is proven correct experimentally.

It appears that the confused world has accepted it as both physics and religion.

At issue for mechanical physics or natural science is which material particles collide, interacting with matters to produce all universal phenomena, because they should have physical origins. The presence of all universal phenomena strongly suggests the presence of an atmosphere of some kind of material particles to collide, interacting with matters to produce them. They must be so tiny that they have eluded experimental detection so far. However, this is the scientific problem alone to mechanical physics, not that of mathematical physics. Teaching mathematical origins of mainstream-of-thought physics does not want an atmosphere of tiny particles to collide, interacting with matters to produce universal phenomena. In fact, physics community is afraid of such an idea, which will lead to discover the physical origins of universal phenomena leading disproving the entire mainstream-of-thought physics.

As early as the eighteenth century, Le Sage had proposed a gravity theory suggesting that there was an atmospheric tiny particles capable of penetrating through matters to collide, interacting with matters to produce gravity, but those, such as Maxwell, supporting mathematical physics, has unduly discredited his theory. It has been known for more than a half century that there is an atmosphere of neutrinos in the space of the universe constantly produced by the nuclear reactions in all stars. Although neutrinos are very elusive, scientists have been able to detect them indirectly and found that nuclear reactions produce them. Therefore, all stars constantly produce neutrinos from their nuclear reactions to maintain an atmosphere of energetic neutrinos in the universe. The physics community teaches that matters having atomic and molecular structures are made of very small point-sized subatomic particles and inside matters they are far apart from one another; therefore, matters are essentially empty space to neutrinos. So, neutrinos penetrate through matters, essentially hardly having any chance to collide, interacting with their subatomic particles. Therefore, the physics community teaches that the atmospheric neutrinos do not collide, interacting with matters to produce any universal phenomena.

The following arguments have been used by the scientific community to discredit Le Sage's gravity theory have also been used to rule out the possibility that neutrinos are responsible for producing gravity (and other universal phenomena). If effective collision interactions between neutrinos and matters occur to produce gravity, their collisions should be inelastic and simultaneously produce huge amounts of heat. And based on the calculation of Maxwell and others, the heat produced should vaporize matters in seconds. The heat apparently is missing to support Le Sage's theory.

It further points to the fact that the planets orbiting the sun don't slow down to prove that the atmosphere of neutrinos does not produce drag effect or collide interacting with matters to slow down them.

Mechanical physics assumes that nature has infinitely large space and any gas-like matters such as neutrinos in our universe should continue to disperse or escape out of the universe to dilute to eventually nonexistence. To constantly have an atmosphere of neutrinos in the universe, there must be many sources about being evenly distributed in the space of the universe to constantly produce and emit them. It has been known that the nuclear reactions inside all stars produce neutrinos to maintain an atmosphere of neutrinos in the universe.

It is also known that neutrinos are uniquely capable of penetrating through all matters having atomic and molecular structures; therefore, the atmospheric neutrinos are the only particles taking up all the space of universe, including the space taken up by matters as well. They, therefore, should be the only possible particles in the space of the universe to collide, interacting with matters to produce all the universal phenomena. Yet the physics community has been unwilling to seriously look into this scientific possibility both due to scientific misconceptions and the protection of its mainstream-of-thought physics.

It is well-established scientific knowledge that the collision interactions among matters produce physical, chemical, and nuclear reactions, which simultaneously release both light and heat, and the collision interactions cause the orbital electrons of both atoms and molecules to emit or absorb light. However, even to date, scientists do not really know what light and heat are and how orbital electrons emit them. Also, collision interactions can make atomic nuclei radioactive, even leading to nuclear reactions releasing intense light and heat besides subatomic particles. Atomic nuclei are known to be made up with protons and electrons and again it has not been understandable why nuclear reactions release intense light and heat. These findings however strongly suggest that (orbital) electrons, atomic nuclei, and maybe even protons contain light, heat, and neutrinos, but surprisingly, the scientific community has so far ignored this possible scientific interpretation.

It is also the well-established scientific knowledge that there are two elementary charge particles, proton and electron, making up all matters and the flow of electrons in metals produces electricity, which has played a major role to create today's civilization. Yet, the scientific community has no scientific understanding what the charge energy both protons and electrons constantly carry is and where it comes from. Although the physics community has mathematically interpreted all universal phenomena, including the charge energy of electrons and protons, it has so far ignored the science of what charge energy is and why and how proton and electrons constantly carry it. Mechanical physics teaches that collision interactions produce phenomena; therefore, the collision of neutrinos with proton and electron particles should logically produce charge energy. The only scientific possibility is that for some scientific reason proton and electron can effectively collide, interacting with atmospheric neutrinos to produce their charge energy.

One possibility is that a dense cloud of neutrinos constantly surrounds both electrons and protons. This possibility makes sense because mechanical physics teaches that the collision interactions of material particles produce energies and other phenomena. A proton or an electron, therefore, is likely made of a large, electrically neutral particle surrounded by a dense cloud of much smaller particles, which should be the neutrinos constantly supplied by the atmospheric neutrinos of the universe. This only possible scientific interpretation of the charge energy of both electrons and protons can also be applied to explain how the atmospheric neutrinos can effectively collide, interacting with matters to produce all universal phenomena. Therefore, neutrinos are likely the common physical origin of all universal phenomena.

The collision interactions between atmospheric neutrinos and protons, electrons, or matters are mainly those among dense neutrinos clouds and atmospheric neutrinos, which are mainly collisions among neutrinos and so far have not been expected and are unknown to the scientific community. It will explain in chapter five why both electrons and protons are surrounded by a dense, turbulent dust cloud of neutrinos (and ethers) and how their charge energy is produced. These unexpected collision interactions produce the charge energy both protons and electrons constantly carry, leading to the formation of matters having atomic and molecular structures, producing gravitational force, other universal forces, and universal phenomena.

However, unlike the collision interactions among matters familiar to the scientific community, they do not lead to atomic and molecular motions and are therefore ineffective of producing light and heat. So, the scientific community has been wrong to expect that producing gravity should also produce large amounts of heat to unduly discredit Le Sage's gravity theory.

The scientific interpretation of charge energy has further led to the fundamental understanding of all types of collision interactions, particle physics, the development of a new cosmology showing off the unique beauty of the consistency, coherency, and simple logical scientific understanding of mechanical physics and natural science.

Over the past century to support its own mainstream-of-thought physics, the scientific community has led Einstein's scientific revolution against mechanical physics or natural science. For example, it teaches mathematical origins and makes physical origins taught by mechanical physics unacceptable for use in interpretation of universal phenomena. It has also rejected Le Sage's gravity theory and the scientific possibilities that neutrinos may somehow collide effectively with matters to produce universal phenomena.

It is unlikely that neutrinos can knock any electron or proton off any atom and/or molecule; therefore, the neutrino detection method based on this scientific principle may not really work for the detection and study of neutrinos. It is possible that stars and exploding stars also emit trace amounts of particles, which are much larger than neutrinos capable of knocking subatomic particles off atoms and molecules, and the scientific community has mistakenly studied them as neutrinos.

A major scientific R&D effort is to study neutrinos emitted from the sun. However, the study methods used may have been studying those rare particles much larger than neutrinos. The findings so obtained may have led to the misunderstanding of neutrinos by the scientific community. Both wrong scientific concepts about neutrinos (and ethers) and erroneous experimental methods used may have prevented the progress of the R&D of neutrinos and related scientific topics over half a century.

Earth, all other planets, and their moons in the solar system appear to have steady-state orbits. These findings have become the strongest scientific evidence for the physics community to conclude that the atmosphere of neutrinos (and ethers) has no drag on matters in motion. However, a close examination of the solar system reveals much orderliness, which cannot occur randomly, hinting that there are forces other than gravitational force to shape up the solar system and other similar heavenly systems.

First of all, the fact that all planets circulate the sun in counterclockwise direction suggests that there may be some kind of force supporting this type of planetary motion against their clockwise orbiting motion. If all the planets have been captured randomly one at a time by the sun, as the mechanical physics-based cosmology given in chapter six teaches, all the planets caught by the sun circulating clockwise, moving against this force, must have plunged into the sun. Besides, this unknown force should also be responsible for (gradually) bringing all the planets to almost on the same plane.

The orderliness of the solar system therefore tells us that something inside the sun constantly produces a force in the direction all its planets circulate the sun and counter its gravity force as well. The newly discovered cosmology by the author teaches that all bright stars have a "dense-matter object," which is an antigravity matter, at their center undergoing nuclear reactions on its surface only. Some rotate very quickly to produce a spiral wind of neutrinos and ether and other particles produced by nuclear reactions, which cause both the stars to rotate and their planets circulating in the direction the dense-matter object rotates to have pseudo steady-state orbits. Please read chapter six for details.

Physics is based on elastic collision interactions among particles carrying no or negligible amount of charge energy, obeying Newton's three laws of motions, while chemistry is based on inelastic collision interactions among particles carrying charge energy, which disobey Newton's laws of motion. The teaching suggests that there may be a fundamental contradiction between physics and chemistry, which has not been addressed so far. The author found that the charge energy of both protons and electrons, which lead to inelastic collision interactions, can be interpreted by the combination of two opposite elastic collision interactions: the pushing force of the dense turbulent dust cloud of neutrinos of protons or electrons and the atmospheric pressure of the atmosphere of neutrinos of the universe. He therefore concludes that

- all collision interactions are elastic; and
- an inelastic collision interaction is the combined effects of more than one elastic collision interactions.

Therefore, there is no scientific contradiction between chemistry and physics, which should not be allowed to exist in science. Of course, it is understandable that inelastic collision interactions also involve structural breaking up such as breaking physical and chemical bonds. Therefore, physics and chemistry are consistent and coherent as expected that natural science should be both consistent and coherent throughout all fields.

It is a mechanical universe with the very existence of its matters from their constant collision interactions. Both protons and electrons constantly collide, interacting with atmospheric neutrinos (and ethers) to gain energy to maintain their charge energy needed to keep their atomic and molecular structures. All matters having atomic and molecular structures are therefore in dynamic equilibrium with the atmospheric neutrinos (and ethers) constantly getting their supply of kinetic energy in order to continue to exist. The amazingly comprehensive scientific view is that it is a universe of neutrinos and ethers. Everything in the universe needs to breathe the atmospheric neutrinos and ethers to exist. Without them, everything will disintegrate and disappear. What we see and touch are all neutrinos and ethers because they surround electrons, protons, and all matters.

Author's New Ether Theory

The author's new ether theory is still unknown to the scientific community. Ether has long been a very controversial scientific topic associated with many wrong scientific concepts. The long controversy debates, however, suggest ether or something like ether has been needed for scientific interpretation of some universal phenomena of light.

Earlier, the author thought that neutrinos were ethers. However, with more scientific evidences available to him, he revised his concepts of ether and light, leading to the development of his new ether theory, which now is an important part of his overall scientific discovery. It offers a logical interpretation of light and a consistent and coherent interpretation of all lights' interactive phenomena, even including the quantum effect of light spectrum. It further suggests an exciting scientific possibility to interpret the complex properties of the microenvironment in terms of mini atoms and mini molecules made of neutrinos and ethers.

The new ether theory is needed to interpret all light phenomena and light itself. Findings have concluded that the collision interactions of matters cause the orbital electrons of both atoms and molecules to emit or absorb light; therefore, they should be neutrinos, which should be invisible, essentially undetectable, and be able to penetrable through matters, but the properties of light and heat are very different from neutrinos. Therefore, they cannot be neutrinos.

Also, findings show that the energy of light reduces slowly while traveling in space to produce the so-called universal redshift phenomenon. Light also undergoes the so-called gravitational redshift and gravitational bending phenomena. Emitted from orbital electrons, light should be neutrinos or something made of neutrinos because it should be emitted from the dense clouds surrounding the orbital electrons. The particle sizes of light, therefore, should be comparable to those of neutrinos, and the collision of light with neutrinos should result in abrupt change in direction and even be destructive.

However, the above-mentioned phenomena show collision interactions that cause gradual change in light energy and/or direction, suggesting that, besides neutrinos, there is an atmosphere of particles having sizes even much smaller than neutrinos in space to collide, interacting with light to produce these phenomena.

The author therefore proposed his new ether theory, suggesting the presence of an atmosphere of particles having sizes even much smaller than neutrinos and light particles in space to collide, interacting with light particles to produce the all light-interactive phenomena, and he calls these particles "ethers" because they play the expected scientific roles of ether well. To be constantly present in space, ethers should also be produced by the nuclear reactions in all stars simultaneously with the producing of neutrinos. In the presence of an atmosphere of ethers, individual neutrinos may be surrounded by a dense, turbulent cloud of ethers, similar to that both protons and electrons are surrounded by a dense turbulent dust-cloud of neutrinos (and ethers) to produce charge energy, and thus may carry "mini-charge energy" similar to electrons and protons.

If neutrinos have a broad range of particle size distribution, for example, having particle size differences close to that between an electron and a proton, it is possible that mini atoms and mini molecules of neutrinos and ethers may form and commonly exist in the microenvironment, making it very complicated scientifically. The presence of an atmosphere of ethers therefore offers an explanation what light may be: the mini atoms and mini molecules of neutrinos and ethers, which therefore have properties different from neutrinos and show quantum effect as well.

Whether the atmospheric neutrinos and ethers in the universe effectively collide, interacting with matters to produce all universal phenomena, is such an important scientific topic that it determines whether mechanical physics or mainstream-of-thought physics is the correct natural science.

The presence of an atmosphere of energetic neutrinos (and ethers) in space is an undeniable scientific fact. However, if neutrinos and ethers are responsible for producing universal phenomena and are their common physical origin, their presence has disproved the entire mathematical physics or mainstream-of-thought physics. Therefore, the physics community has been teaching and insisting that neutrinos are very inactive and do not collide, interacting with matters to produce any universal phenomena, and there is no ether in space.

From a mechanical physics point of view, it is a well-established scientific fact that

- matters from their collision interactions produce phenomena; and
- all universal phenomena should also be produced by collision interactions between matters and the tiny particles taking up all space.

Because neutrinos and ethers are the only matters in the form of fast-moving, tiny particles in the universe taking up all the space of the universe, including the space inside all matters having atomic and molecular structures, they have to be responsible for colliding, interacting with matters to produce universal phenomena occurring anywhere in the universe such as gravity, the charge energy of protons and electrons, and so forth. The finding of many universal phenomena therefore proves that they are produced by neutrinos and ethers colliding, interacting effectively with matters having atomic and molecular structures to produce them.

Because the atmospheric neutrinos and ethers are the only elementary tiny particles taking up all the space of the universe to produce all universal phenomena, they should be the common physical origin of all universal phenomena. This logical conclusion has greatly simplified scientific understanding of all universal phenomena and the universe, making it possible to have a broad-scope consistent and coherent with logical scientific understanding of nature.

Mechanical Physics-Based Cosmology

It happens that the author's new ether theory disproves the foundation science of the big bang theory by offering an alternative scientific interpretation of the phenomenon of universal redshift of starlight supported by overwhelmingly more findings. Furthermore, the presence of many universal phenomena proving that the atmospheric neutrinos and ethers collide effectively with matters to produce all of them and the finding of their common physical origin to logically consistently and coherently interpret them have disproved the entire mainstream-of-thought physics, including the big bang theory. For example, the atmospheric neutrinos and ethers produce pushing gravity force instead of curved spacetime.

Modern cosmology has been developed based on both Einstein's general theory of relativity and the big bang theory. The big bang theory teaches that the universe was born from a big explosion of a "singularity" about fifteen billion years ago and has been uniformly expanding ever since. Einstein's field equation represents the universe-like electromagnetic wave representing light, and the big bang theory has been included in it by a cosmic constant to the field equation. Einstein postulated that matters somehow bend space to produce gravity, meaning there is something having no mass in space having a mathematical origin called spacetime to produce gravity.

Modern cosmology teaches that the motions of all matters—such as all stars are controlled by gravitational force produced by all matters in the universe and the future of the universe to expand forever to slow down, expanding to reach its maximum size or even beginning to contract to finally collapse—relies on the total matter content of the universe obeying Einstein's field equation.

Over the past century, both Einstein's theory of relativity and the big bang theory have shaken the world and been warmly accepted by the scientific community worldwide despite that they are still scientifically unproven theories. Modern cosmology has been the most popular science for decades, carrying strong religious flavor.

Mechanical physics teaches that the only scientific possibility is that the atmospheric neutrinos and ethers somehow collide effectively with matters to produce gravity and other universal phenomena. Besides, it also means that the wind of the atmospheric neutrinos and ethers pushes stars, their planets, and moons around; therefore, gravity is not the only natural force controlling the movement of stars and planets. For example, the spiral wind of the atmospheric neutrinos and ethers, not gravitational force, likely control the circular orbital motions of the stars inside a spiral galaxy.

The scientific community has found that the amount of matters inside a universe is too small to produce strong enough gravitational force to keep the spiral movement of its stars without flying apart. The scientific community therefore has postulated that most matters are undetectable dark matters and dark energy to account for the unexpectedly strong gravity.

The new mechanical-physics based cosmology developed by the author is based on R&D findings at very close to absolute zero temperature, showing the atomic orbital begins to collapse. He postulated the presence of a very dense matter and called it "dense-matter object," having a structure and mass density similar to atomic nuclei, except having equal number of protons and electrons only existing in places at absolute zero temperature and below it. Absolute zero temperature is likely the transition point between matters having atomic and molecular structures and dense-matter objects. At absolute zero and lower temperatures, matters having atomic and molecular structures collapse to form dense-matter objects, which is stable at and below absolute zero temperature.

He reasons that, outside and leaving the universe, the temperature will continue to drop to below absolute temperature, where heavenly bodies will collapse to form dense-matter objects and may agglomerate to grow in sizes. However, like some unstable atomic nuclei, inside the universe under constant bombardment of energetic neutrinos and ethers, dense-matter objects are unstable, becoming radioactive and undergoing nuclear reactions but only at their surface because they have such amazingly dense mass density that even neutrinos and ethers cannot penetrate deep into them.

The author therefore suggests that dense-matter objects are likely the matter that nature is constantly making and storing for making new galaxies and universes in places outside universe having temperatures at and below absolute zero degrees. Because matters having atomic and molecular structures such as hydrogen and helium cannot survive outside universe, they cannot be the raw materials to form baby stars to give birth to a new galaxy or universe.

He further suggests that a violent collision of two large dense-matter objects would produce a galaxy having billions of broken pieces of dense-matter objects, instantly turning them into newborn baby stars starting nuclear reactions on their surface, a perfect sustained-release mechanism of the nature to utilize the nuclear energy of dense-matter objects, making a universe to have long-life stars.

The author has therefore proposed a cosmology model completely different from the modern cosmology we have been learning in the past century. According to him, producing a galaxy is by far a natural phenomenon of violent collision of two dense-matter objects while a universe is made of many galaxies close to one another. There may be many galaxies and universes coexisting in nature or at different time periods. The raw material to make a universe is dense-matter objects not hydrogen and helium. A baby galaxy therefore contains billions of pieces of 100 percent dense-matter objects as baby stars. The most of cosmic hydrogen and helium are produced by the nuclear reactions of the dust particles of dense-matter objects produced by the violent collision of two dense-matter objects.

Inside a newly born galaxy, dense-matter objects undergo nuclear reactions on their surface constantly emitting neutrinos, ethers, and other particles produced by the nuclear reactions, such as hydrogen, helium, and their ions, making the baby stars have strong antigravity. It means that, inside the universe, dense-matter objects or the densest stars such as neutron stars and black holes are dense-matter objects or antigravity matters having strong antigravity instead of strong gravity, as the modern cosmology teaches.

The logics of the author's new cosmology are that a galaxy is born to contain 100 percent dense-matter objects as baby stars, which undergo nuclear reactions to make matters having atomic and molecular structures, which constantly get energy from neutrinos and ethers to produce gravity, and are therefore gravity matters. When its dense-matter objects are used up, the galaxy dies. During the process, a baby star constantly produces hydrogen and helium and throws them far out to eventually form a large ball-shaped outer layer, which gets brighter upon getting thicker and by the heat of the nuclear energy of the dense-matter object at its hollow center to become a regular, big, bright star.

A regular bright star therefore is very large, having artificially low density, and is turning from an antigravity star to a gravity star. A mature regular bright star has a thick outer layer. Constantly heated by nuclear energy, the inner surface of the outer layer of a mature star gets so hot that it begins to undergo nuclear reactions to make elements heavier than helium and gets older quickly.

Therefore, the inner surface of the outer layer of a mature regular bright star such as our sun gets harder with aging, becoming more impenetrable to the gases constantly produced by its dense-matter object. Mature regular stars therefore need to erupt periodically, like sunspots and flares.

It is known that large bright stars undergo supernova explosion to blow away all or part of their outer layer to form such dense stars as neutron stars and white dwarfs. However, according to the author, dense stars are not dead stars, and they are entirely or partially expose dense-matter objects constantly undergoing nuclear reactions. Both black holes and neutron stars are dense-matter objects, the densest stars. They are also antigravity stars having strong antigravity instead of having such strong gravity that light cannot escape, as the physics community teaches. Large dense-matter objects look like black holes because their antigravity particles destroy the light in their surroundings. White dwarfs are the dense-matter objects still having a thin outer layer, which is very close to their dense-matter objects and are therefore very hot. Their outer layers should continue to undergo nuclear reactions, making even heavier elements become even harder for the gases produced by nuclear reaction to penetrate through.

White dwarfs are known to undergo the most violent supernova. As expected, the mechanical physics-based cosmology is still the simple three-dimensional universe logically explained by findings such as what are the matters to form galaxies, universes, stars, planets, and so forth and the forces to control their movements. Please read chapter six for details.

The Motions of Solar System Logically Explained

The steady-state orbit and rotational motions of planets and their moons have long led the scientific community to teach that there is vacuum space in the universe and gravitational force is the only universal force controlling the motions of stars and their planets in solar systems, galaxies, and universes. Upon the discovery of an atmosphere of neutrinos in space, the scientific community teaches that matters having atomic and molecular structures are made of point-sized, tiny subatomic particles (protons and electrons) far apart from one another. Therefore, matters are essentially vacuum space for neutrinos to freely penetrate through them. Atmospheric neutrinos therefore have no drag effect on moving matters, such as planets.

It is, however, very strange that all planets orbit the sun in the same counterclockwise direction and even essentially on the same plane. In addition, the sun is also rotating in the same direction the planets are orbiting. Together, all these motions point to a possibility that, besides gravitational force, something inside the sun constantly produces a force to push the sun to rotate and even to help all planets to counter both the drag force of atmospheric neutrinos and ethers and the sun's gravitational force, allowing them to have steady-state orbits. Apparently, it is a specific directional force to push all planets to orbit in the direction the sun rotates. It means that, when the sun catches a planet orbiting in the wrong direction, such as clockwise, it will quickly plunge into the sun. Statistically speaking, around ten such planets have plunged into the sun already.

According to the mechanical physics-based new cosmology, a regular bright star like our sun is made of a hollow, ball-shaped outer layer containing mainly hydrogen and helium having a dense-matter object at its center. Because nuclear reactions are constantly ongoing at the surface of dense-matter objects and many of them have irregular shape, thus, are fast-rotating and even doing fast circular motions. Astronomers have already found that many neutron stars, which are dense-matter objects, are fast-rotating and call them "pulsar." Therefore, many dense-matter objects inside stars such as our sun are also fast-rotating, throwing out neutrinos, ethers, hydrogen and helium gases, and their ions out as a spiral wind storm. They are likely the source of the forces responsible for the motions of the solar system. With the energetic particles it constantly emits, it should produce a pushing force to make the outer layer of the bright star to rotate. Meanwhile, both the dense-matter object and the rotating outer layer of the sun should create a spiral wind of neutrinos and ethers surrounding the sun to offer its planets a directional force to support their steady-state orbit and rotational motions.

When planets are caught (one at a time) orbiting in the direction the dense-matter object is rotating, it will survive, but its orbit will be gradually adjusted to the very direction the dense-matter object is rotating, resulting in all planets orbiting essentially on the same plane.

As expected both the atmospheric neutrinos and ethers do collide, interacting effectively with matters to produce drag effect, gravitational force, and other universal phenomena. The steady-state orbit and rotation motions of the planets and the orderliness of the solar system further prove that the sun's dense-matter object is an antigravity matter and is fast-rotating and even fast circulating. Therefore, its antigravity force in the direction of rotation and circulation has not been entirely canceled by its gravitational force, thus allowing the sun to have planets and moons having pseudo steady-state orbits.

It is likely that dense-matter objects have their own specific speeds of rotation and circular motions; therefore, all stars may have their own unique planets supporting characters. Many stars with their dense-matter objects not rotating or slow-rotating likely cannot support pseudo steady-state planetary motions and therefore have no planet.

Matters-Energies Not Mutually Convertible

The famous Einstein equation has long been nicknamed the "God equation," teaching that matters and energies are mutually convertible. Einstein mathematically derived it based on postulations that light was electromagnetic wave having no mass and an absolute constant speed and nothing else can reach light speed. The physics community has long been teaching that matters and energies are mutual convertible, obeying the Einstein equation. Nuclear reactions have been found to produce huge amounts of energies upon losing tiny amount of mass. These findings have been used to prove that mass turns into energy, obeying Einstein's equation. Therefore, the physics community claims that Einstein's equation has been proven correct experimentally.

However, it is well known that nuclear reactions are chain reactions, which are very difficult to control. They usually happen together with many chemical and physical reactions. It is doubtful that, in a nuclear reaction, the tiny amount of the matters converting to energy can be accurately determined and Einstein's equation can be quantitatively proved experimentally.

The major scientific problem is that pure energy or mass-free energy described by Einstein's equation is scientifically indefinable. Postulating the presence of particles and/or energies having no mass by the physics community has been a major step leading science into pseudo-science because any phenomenon detectable experimentally should be produced by matters, which have mass. Light is detectable, proving that light has mass and disproving the fundamental postulation of Einstein's theories of relativity that light is electromagnetic wave having no mass.

Findings that atomic nuclei undergo nuclear reactions emitting neutrinos should prove that atomic nuclei contain neutrinos (and ethers) but not that neutrinos and ethers are somehow created by nuclear reactions. It further means that both electron and proton contain neutrinos and ethers as well. Now, both electrons and protons contain neutrinos and ethers have also been proven by that they both constantly carry charge energy. Therefore the findings that nuclear reactions lose small amounts of masses and produce huge quantities of energy can be interpreted by that nuclear reactions release large amounts of neutrinos and ethers to produce huge sums of energy instead of converting mass to pure energy. Therefore, the small amounts of masses lost in nuclear reactions prove that nuclei have released large quantities of neutrinos and ethers in the forms of huge amounts of light and heat, not that the lost masses or matters have been converted to pure energy. Therefore, Einstein or God equation has no scientific meaning, and matters have not been proven convertible to pure energy.

Chapter 5 – Mechanical Universal Phenomena

Abstract and Introduction

The only possible scientific interpretation of the charge energy of both proton and electron constantly carrying is that the non-charge particle of both of them are surrounded by a dense, turbulent dust-cloud of neutrinos and ethers coming accumulated from the atmospheric neutrinos and ethers constantly produced by the nuclear reactions occurring in all stars. Neutrinos are much smaller than the non-charge elementary particles of protons and electrons, and ethers are even much smaller than neutrinos. The author postulated the existence of an atmosphere of ethers to interpret many light-interactive phenomena such as universal redshift of starlight.

The charge energy of protons and electrons is the combined effects of the atmospheric pressure of neutrinos and ethers on both electrons and protons and the pushing force among the dense, turbulent dust-clouds surrounding both electrons and protons. Atmospheric neutrinos and ethers carry the huge energy of nuclear reactions to constantly collide, interacting with the dense dust-cloud particles of all protons and electrons to constantly supply them energy needed to constantly carry charge energy, to form atoms and molecules, and to produce other universal phenomena.

The above scientific understanding of charge energy leads to the understanding that neutrinos and ethers are the common physical origins of all universal phenomena and a broad scope scientific understanding for all of them. It also leads to the understanding that the charge energy of both proton and electron is produced by the combined effects of two opposite elastic collision interactions. Therefore, the effect of an inelastic collision interaction is the combined effects of two or more elastic collision interactions. Chemistry, the result of inelastic collisions, therefore is consistent with physics, the result of elastic collision interactions, with both obeying the laws of elastic collisions such as Newton's laws of motion. It further affirms that natural science is consistent and coherent throughout all fields.

Upon discovering mechanical-physics based interpretation of universal phenomena, particle physics, cosmology, mechanical physics has been proven to be the physics of everything and natural science the only correct or real science. As a result, the entire physics of attraction forces acting from distance includes the entire mainstream-of-thought physics has being scientifically disproved.

This book broadly covers scientific topics in physics, cosmology, and chemistry. Wikipedia broadly covers the historic background knowledge and is therefore a good review reference for this book. Chemistry studies all matters having atomic and molecular structures found on earth and likely also existed in the entire universe. Both atoms and molecules are made of electrons and protons, the only two stable fundamental charge particles making up all matters having atomic and molecular structures. The charge energy constantly carried by both electrons and protons is responsible for their collision interactions to form atoms and molecules; therefore, scientific understanding of their charge energy is of fundamental importance.

There are also electrically neutral or non-charged particles such as steel balls and inert gases. The collision interactions among them obey Newton's three laws of motion[1] also known to be the laws of "elastic" collisions, which do not result in combining them together to form their "compounds," like what charge particles do. Matters made of atoms and molecules are formed by electrons and protons from their "inelastic" collision interactions. The physics laws of motion therefore appear only to govern non-charge particles or when their charge energy is negligibly small and maybe neutrinos,[2] the smallest none charge particles known.

Relying on inelastic collision interactions of electrons and protons to form all matters, chemistry appears contradictory with physics. However, contradiction should not be allowed in the entirety of natural science, and a scientific reason should exist to explain it. Although mathematics-based interpretation and quantification of both charge energy and their forces have long been given, these are based on both the findings of the existence of these charge particles and their properties without scientific understanding of the charge energy and their energy source.

The physics has essentially been split in two since the eighteenth century by Newton's theory of universal gravitation: the physics of pushing forces acting on contact and the physics of (attraction) forces acting from distance. The former is known as the classical or mechanical physics, and it teaches that matters produce phenomena, meaning that phenomena have physical origins. For example, air is the common physical origin of atmospheric pressure, sound, wind, and so forth. The latter physics has led to the development of today's mainstream-of-thought physics teaching that universal phenomena have mathematical origins such that light is electromagnetic wave or photon and gravity is curved spacetime and they all have no mass.

Based on mechanical physics, many scientists, such as Newton[3]

---

[1] I. Newton, *Philosophia Naturalis Mathematica* (1687).

[2] F. E. Close, *Neutrino* (Oxford University Press, 2012).

[3] I. Newton, *Opticks* (1704).

and Le Sage[4] have long predicted that tiny particles produce universal phenomena such as light and gravity. Le Sage further postulated that the tiny particles producing gravity were capable of penetrating through matters to collide interacting with all of their (subatomic) particles to produce gravitational force having strength proportional to the mass content of matters.

Modern knowledge teaches that the nuclear reactions inside all stars continue to produce and to emit neutrinos, therefore maintaining an atmosphere of energetic neutrinos inside the universe, and they are capable of penetrating through matters. Neutrinos are likely the tiny particles Le Sage postulated to produce gravity. However, the physics community has long discredited his gravity theory[5], arguing that, to produce gravity, the collision interactions between neutrinos and matters should produce huge amounts of heat capable of evaporating matters in seconds, but the heat is missing.

Besides, physicists continue to teach that protons and electrons are tiny point-sized particles and, inside atoms and molecules, they are separated far apart from one another. Therefore, the physics community teaches that to neutrinos, matters having atomic and molecular structures are essentially empty space. And therefore neutrinos penetrate through matters, essentially freely and hardly having any chance to collide, interacting with their protons and electrons, thus, not responsible for producing any universal phenomenon.

The scientific community has long been teaching that a proton carries a positive charge while an electron carries a negative charge, enabling them to collide interacting with one another to form atoms and molecules. The teaching however cannot explain why the negatively charged orbital electrons in atoms and molecules do not simply fall into the positively charged atomic nuclei. Many scientists believe and teach that the charge energy of both electrons and protons are their spin energies, but they also cannot explain

[4] Georges-Louis Le Sage, *Memoires de l'Academie Yoyale des Sciences et Belles de Berlin* (Lucrece Neutonien, 1784), 404–432.

[5] J. C. Maxwell, *Encyclopedia Britannica*, 9th ed., s.v. "atom."

how spinning particles produce collective directional charge forces and where both electrons and protons constantly get their spin or charge energy from.

In the past century, the physics community has been teaching both mechanical physics and mainstream-of-thought physics despite the fact that the two physics contradict with each other. Before Einstein's scientific revolution, mechanical physics had long been repeatedly proven correct, the fundamental science of all other fields of natural science, and the daily applied physics.

This book offers the breakthrough scientific discovery of the author using mechanical physics to interpret those scientific topics already interpreted by mainstream-of-thought physics including universal phenomena, cosmology, particle physics, and more against the claims of the physics community that it is impossible to do so. The breakthrough scientific discoveries therefore prove that mechanical physics is the physics of everything, meaning that natural science is the only correct or real science.

A New Ether Theory

Light, including heat, has been the most studied, technologically applied, and utilized phenomenon, but it is still the least scientifically understood and the most controversial scientific topic. Historically, one of the most controversial scientific debates has been whether there is an atmosphere of ether to collide interacting with light such as assuming transmitting light. Based on mechanical physics, lights should be tiny particles, not matter-free waves. And unlike waves interfering with each other in space, tiny particles hardly collide or interfere with one another. There are, however, several phenomena such as the universal redshift of star light, gravitational redshift, and gravitational bending of light, indicating that light traveling in space does collide, interacting with some kind of particles, which should be even much smaller than light particles.

The author has therefore proposed a new ether theory suggesting there is an atmosphere of ethers or particles even much smaller than light particles and neutrinos, which are likely the main component of light particles. To constantly have an atmosphere in the universe ethers should be produced by the nuclear reactions in all stars simultaneously with neutrinos. Similar to neutrinos and maybe even more so, the concentration of ethers inside and around matters is much higher than that in space. This explains why the effects of both gravitational redshift and bending of light are much stronger effects than the universal redshift of starlight, which takes millions of light-years to develop gradually.

Capable of interpreting all the above-mentioned three light-interactive phenomena consistently instead of only one, the new ether theory is more likely the correct theory to interpret the phenomenon of universal redshift of starlight rather than the big bang theory. Besides, the new ether theory can also explain many other universal phenomena such as the quantum effect of light and even what light particles are. Therefore, new ether theory has disproved the big bang theory, which happens to be the foundation science of modern cosmology. The breakthrough discovery of a mechanical physics-based new cosmology given in chapter six has further disproved the big bang theory independently.

The Collision Interactions of Atmospheric Neutrinos and Ethers with Matters: To Produce Universal Phenomena

Introduce a new type of particle collision interactions still unknown, unexpected, and unacceptable to the scientific community. It is the type of collision interactions among atmospheric neutrinos and ethers and matters. It has not been expected and considered possible for this type of collision interactions to occur effectively.

The author found that the only scientific possibility to interpret the charge energy of both proton and electron constantly carrying was that the non-charge elementary particle of both electron and proton are surrounded by a dense dust-cloud of neutrinos and ethers to collide interacting with atmospheric neutrinos and ethers to produce the charge energy. They have to come accumulated from the atmospheric neutrinos and ethers constantly produced by the nuclear reactions occurring in all stars. Among them neutrinos are much smaller than the non-charge elementary particles of protons and electrons and ethers are even much smaller than neutrinos.

The above only possible interpretation suggests that both proton and electron have high contents of both neutrinos and ethers and that matters having atomic and molecular structures are occupied by dense dust-cloud of neutrinos and ethers not having essentially empty space as taught by the scientific community. It also offers a natural scientific mechanism the atmospheric neutrinos and ethers to collide interacting effectively with matters to produce universal phenomena. The new scientific concept therefore has opened the secret door of the nature for the author to understand why atmospheric neutrinos and ethers can collide interacting effectively to produce all the universal phenomena including gravity, light, charge energy, and so forth, whose scientific secrecy have puzzled the scientific community for many centuries already. It turns out that these universal phenomena have a common physical origin, the atmospheric neutrinos and ethers, and therefore can be interpreted logically, consistently, coherently to further advance scientific understanding of the nature. Although the atmospheric neutrinos and ethers are the only possible particles to collide, interacting with matters to produce these universal phenomena, discovering their scientific roles would scientifically disprove the entire mainstream-of-thought physics developed, supported, and taught by the physics community in the past century. Clearly, physics community has made every effort to disallow looking into this scientific possibility in the past century.

There is a scientific logic behind the above scientific interpretation: the energetic atmospheric neutrinos and ethers collide, interacting effectively with matters to constantly supply them with kinetic energy resulting in producing all universal phenomena. Contrary to the teachings of the physics community that the space inside matters is essentially vacuum, it instead is filled with very high concentration of neutrinos and ethers as evidenced by that both electrons and protons constantly carry charge energy and the presence of all the universal phenomena.

This type of collision interactions mainly are those among neutrinos and ethers themselves resulting in producing many universal phenomena such as charge energy, universal forces, and other universal phenomena. They are the collision interactions still unexpected, unknown, and even not acceptable particularly by the physics community. They however are not effective in causing atomic and molecular motions and therefore not effective in producing light and heat as wrongly expected by the scientific community from the knowledge of the type of collision interactions among matters scientists are already familiar with.

Producing Charge Energy of Protons and Electrons

Scientists have long discovered many universal phenomena such as light, gravity, other universe forces, and so forth, but they still don't know what they are, how are they produced, and where their energies come from. Mechanical physics teaches that matters produce phenomena; therefore, matters or particles should produce universal phenomena. Le Sage's gravity theory postulates that tiny particles capable of penetrating through matters collide interacting with matters to produce gravitational force. However, physics community has long discredited Le Sage's gravity theory with scientific questions unanswerable at the time, giving itself the excuse to develop and teach mathematical physics. It suggested that the collision interactions of the tiny particles of Le Sage with matters to produce gravity are inelastic and should simultaneously produce large amounts of heat. And according to its calculation, the heat should vaporize matters in seconds and is apparently missing. The physics community further questioned where is the energy source for Le Sage's tiny particles to continue to produce gravity.

Nevertheless, producing gravity by tiny particles colliding interacting with matters has always been the only scientific possibility of mechanical physics. If neutrinos and ethers are the postulated tiny particles of Le Sage's gravity theory, the energy source question should be the nuclear energy of all stars. The missing heat question has been a scientific misconception of the physics community as briefly addressed earlier.

Mechanical physics teaches that particles collide interacting to produce energies or phenomena. Therefore, particles collide interacting should also produce the charge energy that both protons and electrons constantly carry. Since both neutrinos and ethers are present everywhere in the universe including the space taken by all matters, they have to be responsible for colliding interacting with both protons and electrons to produce charge energy. Also, to constantly carry charge energy from the collisions, they must constantly get the kinetic energy of the atmospheric neutrinos and ethers from the collision interactions to produce their charge energy. To be able to effectively collide interacting with atmospheric neutrinos and ethers, both proton and electron are likely made of their non-charge particle surrounded by a dense dust cloud of neutrinos and ethers, which constantly bombarded by incoming energetic neutrinos and ethers should be turbulent.

Having a dense cloud of neutrinos and ethers the incoming atmospheric neutrinos and ethers can effectively collide interacting with the cloud particles of both proton and electron to exchange energy and to maintain their charge energy. Without charge energy, a proton or an electron should be a single non-charge particle much bigger and much more massive than neutrinos.

Judging from the knowledge of atomic nuclei, the non-charged particles of both protons and electrons should have amazingly dense mass, and they likely have about the same mass density as both neutrinos and ethers. Therefore, neutrinos and ethers cannot penetrate through any of the non-charge particles of proton and electron.

The mechanism why a dense dust cloud of neutrinos and ethers is formed around both protons and electrons is yet unclear. It may mean that the collision interactions of neutrinos and ethers with their non-charge particles are somehow "inelastic" to slow down neutrinos and ethers around them. And the atmospheric pressure of neutrinos and ethers should also play an important role for the formation of the dense dust-cloud, which under the constant bombardment from incoming energetic neutrinos and ethers, should be turbulent tending to expand outward.

Therefore, in dynamic equilibrium, there is a dense, turbulent dust-cloud of neutrinos and ethers surrounding the non-charge particle of proton and electron making both of them detectable as carrying charge energy. Therefore, constantly carrying charge energy by both electrons and protons along with the presence of other universal phenomena proves that atmospheric neutrinos and ethers can effectively collide interacting with protons and electrons to produce all of them.

The above scientific interpretation suggests that both electron and proton are not single point-sized particles but composite particles containing besides a non-charge elementary particle, a dense turbulent dust-cloud of neutrinos and ethers surrounding it. Therefore, both proton and electron also contain neutrinos and ethers.

In dynamic equilibrium with the atmospheric neutrinos and ethers of the universe, there are equal number of neutrinos and ethers entering and exiting the turbulent dust clouds of both protons and electrons. Because the entering neutrinos and ethers have faster speed or stronger kinetic energy than the outgoing ones, electrons and protons become the low-energy centers in the atmosphere of neutrinos and ethers of the universe like all planets and some stars. As a result, there is always an atmospheric pressure of neutrinos and ethers on them or a pushing force toward them, which also pushes them toward each other when they are close enough. Meanwhile, the dense, turbulent dust cloud of neutrinos and ethers surrounding both electrons and protons should always have a repulsive force on others.

Neutrinos and ethers inside matters such as the earth are therefore unevenly distributed. There are the atmospheric free neutrinos and ethers of the universe having higher kinetic energy, but the majority should be the dense dust-cloud of neutrinos and ethers of both electrons and protons.

The atmospheric pressure of neutrinos and ethers therefore pushes both proton and electron toward one another like an "attraction force." For convenience, it is sometimes called "attraction force" in this paper. Besides, the dense, turbulent dust-cloud of neutrinos and ethers constantly pushes outward, thus having a repulsive or pushing force on others. Therefore, there are two independent fundamental forces to create the charge energy of both electrons and protons instead of just a single positive or negative force as the scientific community believes and teaches.

It is now understandable why the collision interactions among protons and electrons can form atoms and molecules from the balancing of their two opposite fundamental forces. Also, the so-called inelastic collision interactions now can be explained by multiple opposite elastic collision interactions; therefore, the appearance of the existence of fundamental contradiction between physics and chemistry now can be explained.

The observed charge-energy relationship or "chemistry" among protons, electrons, and their matters given below is unique among them due to the specific size difference among other possible unique collision interaction properties specific among them. Because a proton is much bigger than an electron, the attraction force between them becomes the stronger force than their mutual repulsive force resulting in "attracting" each other initially. However, the repulsive force between them increases more rapidly than their attraction force with their approach to each other, and there is a distance between them where the two fundamental forces are balanced; therefore, an electron will not fall all the way into a proton. The reason that a proton repulses another proton and an electron repulses another electron is that, having the same size, their mutual repulsive force is maximized, which becomes the stronger fundamental force of the two.

Again, this charge-energy relationship and its physics and chemistry are specific to protons, electrons, and matters made of them, which, for example, cannot applied to other "elementary charge particles" found in cosmic ray and produced in high-energy particle accelerators. By doing so, the interpretations based on the above knowledge would be wrong and misleading.

Findings show that free electrons and free protons produce long-range electric and magnetic forces, meaning that their dense, turbulent dust-cloud of neutrinos and ethers collide, interacting effectively with atmospheric neutrinos and ethers, and can continue to extend outward to produce a particle field to produce electric and magnetic forces. There are natural occurring magnets and also ways to make them. The outer orbital electrons of atoms and/or molecules of these magnets also interact strongly with neighbor atomic nuclei. And as a result, they are orderly aligned to have protons partially exposed to become the (+) pole at one end while electrons partially in excess to become the (-) pole at the other end.

Findings have also confirmed that protons and electrons form both atoms and their molecules making up all matters, which are essentially neutral, carrying no significant amount of charge energy. Atoms are made of atomic nuclei densely packed with electrons and protons with protons in extra to be positively charged, and orbital electrons take up very large space around them. A proton attracts an electron by their stronger mutual attraction force. However, the repulsive force between them increases more rapidly upon approaching each other until the two fundamental forces are balanced, and then the electron orbits around the proton while oscillating like a transverse wave to form a hydrogen atom. This should be the mechanism to form all other atoms and molecules as well.

Findings also show that an electron can fall much deeper into a proton in the presence of excess protons to enhance their mutual attraction forces to form atomic nuclei having amazingly dense mass. A neutron is formed by combining an electron and a proton and is stable only inside nuclei in the presence of excess of protons. Many stable atomic nuclei contain proton(s) and about half its number of electron(s). The two-to-one ratio of proton to electron may allow one electron to fall in between two protons, maximizing their attraction interactions and stability.

Atomic nuclei have excessive protons exposed to attract

electrons; therefore, atoms and molecules have orbital electrons. However, the repulsive force among these orbital electrons keeps them far apart and away from atomic nuclei. Under an atmosphere of neutrinos and ethers of the universe and constantly bombarded by these energetic particles, atoms and molecules are in dynamic equilibrium with them constantly getting their kinetic energy to maintain the charge energy of protons and electrons and their bonds inside atomic nuclei, atoms, and molecules keeping them stable.

Findings have also shown that one-to-one proton-to-electron ratios form essentially electrically neutral and stable atoms and molecules. However, all atoms and molecules are not completely electrically neutral, having the so-called dipole moment, which is unique to all atomic and molecular matters for them to have unique physical and chemical properties.

Having the ability to collide interacting effectively with matters to produce the charge energy of both electron and proton means that the collision interactions also result in producing other universal phenomena, which will be further described below.

Producing Gravitational and Other Universal Forces

Capable of effectively collide interacting with matters, the atmospheric neutrinos and ethers entering a planet such as Earth have significantly stronger kinetic energy than those leaving it, resulting in having a wind of neutrinos and ethers blowing with a pushing force constantly toward Earth, which is the gravitational force of Earth. The atmospheric pressure of the neutrinos and ethers of the universe on atomic nuclei should be the strong nuclear force discovered. The presence of an energetic atmospheric neutrinos and ethers also produce the so-called weak force by constantly bombarding all particles such as atomic nuclei and cosmic ray particles, making them unstable. All universal forces are therefore the act-on-contact pushing force of neutrinos and ethers, which therefore are their common physical origin.

The physics community has long been wrong by teaching that universal forces are act-from-distance attraction forces having their own matter-free force fields and force particles to mediate the action of this illogical force.

Because matters having atomic and molecular structures are filled with very concentrated dust-cloud of neutrinos and ethers, they collide, interacting effectively with the incoming atmospheric neutrinos and ethers to continuously gain their nuclear energy, converting it to universal forces and other universal phenomena such as making matters having atomic and molecular structures and producing light. Therefore, this type of collision interaction is responsible for producing most universal phenomena if not all.

Scientific Community teaches that the strength of gravitational force of a heavenly body is proportional to its mass content. However, this is true only for all planets and moons but not true for stars because they are made of antigravity matters or both antigravity and gravity matters. For example, a neutron star is made of an antigravity matter called "dense-matter object" by the author and is therefore an antigravity star. A regular bright star has a dense-matter object at its center and a very large, bright, ball-shaped hollow outer-layer made of matters having atomic and molecular structures, which are gravity matter. Although a regular bright star like our sun is a gravity star, it has been an antigravity star, and its gravitational force increases with aging at the expense of its antigravity matter. Therefore, the gravitational force of a bright star is not proportional to its mass content and not even constant over time.

Light and Theory of Mini Atoms and Mini Molecules

The above scientific interpretations of charge energy and other universal phenomena have left some scientific questions unanswered, particularly regarding the strange behaviors of the dense turbulent dust-cloud of neutrinos and ethers surrounding both electrons and protons. For example, free electrons, free protons, and even partially exposed orbital electrons and protons of atoms and molecules extend their particle field outward to produce long-range electric or magnet forces, but most atoms and molecules are essentially neutral without having significant amount of charge energy. The dense, turbulent dust-cloud of neutrinos and ethers in most atoms and molecules appear to have gained some liquid-like complex properties to restrict their movement, volumes shapes, and so forth, but why?

In addition, well-established knowledge shows that the (outer) orbital electrons of atoms and molecules form complicated chemical and physical bonds, and it is hard to imagine that the dense turbulent dust-clouds of neutrinos and ethers of the orbital electrons can perform such complicated tasks. After all, the most important question what light is remains unanswered.

Findings further show that the collision interactions of matters cause them to emit light and heat, which should be the neutrinos and ethers released from the dense turbulent dust-cloud of neutrinos and ethers of the outermost orbital electrons of atoms and molecules.

Yet, unlike the atmospheric neutrinos and ethers of the universe, light and heat are directly detectable and cannot penetrate deep into matters. Why are the large differences in properties between the neutrinos and ethers emitted from the dense turbulent dust-cloud of neutrinos and ethers of the orbital electrons and the atmospheric neutrinos and ethers of the universe?

The beauty of natural science is its consistency and coherency making it logically predictable. Like an electron and a proton, a neutrino particle also can be surrounded by a dense turbulent dust-cloud of ethers, thus, carrying mini-charge energy. It is also likely that there are different sizes of neutrinos with some having the size difference comparable to that between a proton and an electron. It is therefore possible of forming mini atoms and their mini molecules made of neutrinos and ethers, and they may be commonly present in the universe in the microenvironment.

And it is likely that the dense turbulent dust-clouds of neutrinos and ethers surrounding both electrons and protons offer the best environment for them to form mini atoms and mini molecules. As a result, these dust-clouds have gained very complicated physical and chemical properties to be able to have liquid-like properties, to form close to electrically neutral atoms and molecules, to emit mini atoms and molecules as light and heat, to form different kinds of physical, chemical, and nuclear bonds, and so forth.

Therefore, light (and heat) may be the mini atoms and mini molecules of neutrinos and ethers, thus, having very different properties from individual neutrinos and ethers. Made from individual mini atoms and molecules, light therefore has quantum effect, which happens to be a unique property of light. For example, all universal forces produced by neutrinos and ethers exhibit no quantum effect.

Collision Interactions among Matters

The type of collision interaction among matters is already familiar to the scientific community, which produces light and heat. However, there is still lacking scientific understanding. It is well known to produce physical, chemical, and nuclear reactions emitting or absorbing heat and light. However, the scientific community still does not know what light is and how and why matters emit or absorb light and heat. The scientific understanding of the charge energy of both electrons and protons has further led to the following fundamental scientific understandings of the collision interactions of matters.

Both proton and electron are composite particles, having a dense, turbulent dust-cloud of neutrinos and ethers surrounding the non-charge elementary particle of proton or electron meaning they contain both neutrinos and ethers. It is reasonable to suggest that the collision interactions of matters mainly involve the collision interactions of the (outermost) orbital electrons of atoms and/or molecules resulting in emitting and/or absorbing neutrinos and ethers from their dense dust-clouds of neutrinos and ethers.

However, it is surprising that this type of collision interactions resulting in emitting and absorbing both light and heat instead.

The author has proposed his theory of mini atoms and mini molecules suggesting that the microenvironment of the dense turbulent dust-cloud of electrons inside atoms and molecules is suitable for the formation of mini atoms and mini molecules of neutrinos and ethers. As a result, the properties of these dust-clouds become very complicated difficult to comprehend such as emitting mini atoms and mini molecules as light, forming complex chemical and physical bonds, and so forth. Nuclear reactions also involve emitting neutrinos and ethers in the form of intense light and heat. These and the findings that radioactive nuclei emit neutrinos (and ethers) prove that both electrons and protons or at least one of them contains neutrinos (and ethers), thus, supporting the scientific interpretation of the charge energy of proton and electron.

Because all kinds of reactions (collision interactions) involve emitting or absorbing light and heat or mini atoms and molecules of neutrinos and ethers, they are mass-losing (or mass-gaining) processes, and this conclusion has also been proven by the well-established knowledge of nuclear reactions, which have been proven involving losing small amounts of mass. Therefore, the mass-balance law of all kinds of reactions needs to be modified to take the amount of neutrinos and ethers lost or gained into account. Nuclear reactions should also obey this modified mass-balance law and be included in this scientific law.

In nuclear reactions, the small amount of mass lost is due to their emission of huge amounts of neutrinos and ethers in the forms of intense light and heat. Therefore, nuclear reactions do not create both neutrinos and ethers out of nothing, and there is no need to postulate that pure energy exists and that matters and energy are mutually convertible. Pure energy is not even scientifically definable and Einstein or God equation has no scientific meaning.

Findings further suggest that the content of neutrinos and ethers of an electron (or even a proton) varies with the environment it is in. A static free electron or proton contains the most neutrinos and ethers and is the heaviest. An electron or proton inside the heaviest atomic nucleus contains the least amount of neutrinos and ethers and is the lightest. A positron is likely an electron freshly emitted from an atomic nucleus highly deficient in neutrino and ether content, thus temporarily having properties of "positron" instead of being an antiparticle. Therefore, the existence of antiparticle or antimatter has not been proven to exist.

Surrounded by a dense turbulent dust-cloud of neutrinos and ethers, both electrons and protons should be unstable to motions and accelerations, which should cause the loss of their neutrinos and ethers, thus, their charge energy and mass. For example, in high-energy particle accelerators, the faster they move, the more neutrino and ether, or the more charge energy and mass they lose. Without scientific understanding charge energy, the physics community has misinterpreted the findings of the instability of proton, electron, and other ions to motions and acceleration to support the postulations of Einstein's theories of relativity that the mass of a particle increases with moving speed.

The Fundamental Heat of Matters

All matters made of both atoms and molecules have their protons and electrons in dynamic equilibrium inside atoms and molecules and in between them. These dynamic equilibriums should constantly produce some heat, which the author has called it "the fundamental heat of matters". Matters therefore emit heat at all temperatures. Because transferring heat is a slow process, the center of a matter gets hotter, the larger a matter is. The temperature range of the fundamental heat of a matter also depends on the energy level of surrounding atmospheric neutrinos and ethers. It explains why planets emit more heat than they receive from the sun and why their cores are always hot, having volcanic activities.

It is known that all matters emit light and heat and it is called black-box radiation having a light spectrum characteristics of the temperature range a matter is in. It includes both those produced by atomic and molecular motions and the fundamental heat of matters. Scientists have long found that matters emit all the lights and heat and therefore they should also emit cosmic microwave interference background (CMB). CMB should be mainly the fundamental heat matters emitted in an environment having very uniformly low temperature, approximately 2.7 degrees Kelvin, such as in the intergalactic space surrounding our galaxy, where, without star, is expected to have uniformly low temperature. Therefore, like all other light and heat CMB should be emitted by matters instead of being the residue radiation predicted by the big bang theory.

Chapter 6 - Mechanical Cosmology

A Three-Dimensional Open Universe

Like all other fields of natural science, now the mechanical physics-based cosmology has finally been discovered offering mechanical view of the universe challenging the mathematical interpretation of the universe developed by Einstein and physics community in the past century. If both Einstein's theories of relativity and the big bang theory were scientifically correct, the mechanical view-based cosmology should not exist and never be found. Finding it however proves that mainstream-of-thought physics or mathematics based cosmology taught by the physics community in the past century is scientifically wrong.

As logically expected, it is a three-dimensional universe, where both time and dimensions are presumed absolute but not light speed. There is, however, an atmosphere of neutrinos and ethers constantly produced by stars' nuclear reactions, taking up all space of the universe to collide interacting with all matters to produce universal phenomena.

Contrary to general belief, both the matters and energy of the universe are not conserved because it is an open universe. The stars of the universe constantly undergo nuclear reactions to produce large amounts of energy, neutrinos and ethers, which continuously escape out of the universe, carrying some matters with it as well. For example, some of the outermost stars and even galaxies are leaving the universe.

A universe can be defined as a place where large amounts of stars live. Stars constantly undergo nuclear reactions, releasing huge amount of electrons, protons, neutrinos, ethers, hydrogen, helium, and their ions. The entire universe is therefore a matter- and energy-consuming system having many engines (stars) running constantly burning matter fuel. Because there is always limited amount of fuel for a system to consume, a universe should have a finite duration and size. For a fire to burn, there must be a bigger natural environment to make and to store its matter fuel.

For example, the earth provides an environment for plants to receive light energy from the sun to grow to be stored as the fuel of forest fires, which lightning can start. A forest fire will die when its fuel is used up. The energy-producing material of a universe must first be produced and stored by some nature way before it can be consumed. Also, there should be a nature mechanism to trigger the starting of the consumption of the stored energy-producing material. From what we know, the nature has designed a perfect engine inside all stars to work continuously for probably billions of years maintenance-free to burn the naturally stored energy-producing material by nuclear reactions at essentially constant rates.

A universe therefore needs to be in an even much bigger and longer-lasting natural environment to produce and store energy-producing material and to be born (a mechanism to start consuming the stored material fuel), to grow, and to die in it, which has to be "the domain of the nature" appearing infinitely large and living forever. To understand our universe, we need to figure out some basic properties about the nature, such as what the energy-consuming material the nature produces and stores to create a universe and what mechanism triggers the starting of consuming it or giving birth to a galaxy or universe.

Getting the scientific knowledge about the universe should be a good starting to figure out what the domain of nature outside the universe should be, which appears to have unlimited space and time without the existence of star. There may be many universes inside the domain of nature obeying the same laws of physics and chemistry, existing simultaneously and/or in different time period. A universe is separated from others with much bigger starless space than an intergalactic space. The atmospheric neutrinos and ethers of all universes continuously escape into the huge spaces of the domain of nature. Upon leaving a universe into the vast starless space, the atmospheric neutrinos and ethers continue to reduce their concentrations to lose their kinetic energy gradually. There should still be a diluted low-energy atmosphere of neutrinos and ethers surrounding a universe; therefore, there are forces of the domain of nature similar to the universal forces but much weaker. Both electrons and protons outside universes likely still have weak charge energy. Outside universes, the atmosphere of neutrinos and ethers should be unevenly distributed, and there may be places far away from any universe having close to real vacuum with very low concentration of neutrinos, ethers, electrons, and protons.

Dense-Matter Objects

Both the concentration and energy of the atmospheric neutrinos and ethers continue to decrease upon leaving a universe. The temperature in an intergalactic space is already very low, for example, approximately 2.7 degrees Kelvin, surrounding our galaxy (Milky Way). Upon leaving a universe, the temperature likely continues to drop to 0 degrees Kelvin and even below. Scientists have found that, upon cooling to very close to 0 degrees Kelvin, the (outer) orbital electrons of atoms collapse, and as a result, their mass gets denser, and they have lost their own atomic identity, turning into a new physical state called Bose-Einstein condensate. It is likely that, upon further cooling, their inner atomic orbital electrons will also collapse plunging into their nuclei to form a very dense matter yet unknown to us.

The author therefore has postulated that, at temperatures 0 degrees Kelvin and below in the domain of nature outside the universe, the orbital electrons of matters having atomic and molecular structures collapse to form "dense-matter object," which contains about equal number of protons and electrons. And like atomic nuclei, they have amazingly dense mass. The findings from extremely low-temperature research therefore have suggested that, outside the universe where the temperature likely drops to 0 degrees Kelvin and even below, both atoms and their molecules no longer exist and will collapse to form dense-matter objects. Still under a dilute low-energy atmosphere of neutrinos and ethers or under the weak forces of the domain of nature, dense-matter objects would tend to collect free electrons, protons, neutrinos, and ethers to grow in size. When a dead star or planet collapses to form a dense-matter object, it has an amazingly large mass already. Given time, some dense-matter objects may grow to very large sizes having amazingly large and dense mass and relative kinetic energies.

Under very weak-energy dilute atmosphere of neutrinos and ethers, protons and electrons may still be weakly charged and still have their own dust-cloud of neutrinos and ethers, but it is less turbulent, having smaller repulsion force than those inside a universe. Low-energy neutrinos and ethers are even easier to be caught and trapped by the electrons and protons of dense-matter objects; therefore, they may become heavier, having denser cloud of neutrinos and ethers than those inside the universe. The electrons and protons of dense-matter objects outside the universe therefore may have enriched amounts of neutrinos and ethers, and upon releasing them inside a universe would produce huge amounts of energy. Dense-matter objects therefore are likely the energy-producing matter the domain of nature makes and stores for creating new universes. Therefore, making dense-matter objects, storing and growing in sizes are likely the nature's way for making and storing the energy-producing matter for creating or giving birth to new galaxies and universes.

Dense-matter objects are formed and stable in the domain of nature outside the universe having an environment at and under 0 degrees Kelvin. Having similar structure to atomic nuclei without orbital electrons to protect them, we would expect that, inside the universe under constant bombardment of energetic neutrinos and ethers, dense-matter objects would become unstable, similar to radioactive nuclei. However, because they have amazingly dense mass, even energetic neutrinos and ethers cannot penetrate deep into them, and as a result, only their surfaces are radioactive, releasing neutrinos, ethers, electrons, and protons, along with huge amounts of energy, leading to undergoing constant-rate fusion reactions on their surface to form hydrogen and helium. The nature therefore has the best way to make and store the fuel material outside the universe and use it in a sustained-release way to give long lives to its galaxies and universes.

Inside a universe, dense-matter objects are radioactive, undergoing nuclear reactions constantly shooting out energetic particles to push away all matters; therefore, they are anti-gravity matter, which the scientific community have not yet recognized to exist. In contrast, matters having atomic and molecular structures constantly get energy from the particles emitted by dense-matter objects to produce gravity and other universal phenomena are therefore the gravity matters known to the world.

Because matters having atomic and molecular structures such as hydrogen and helium cannot exist outside universe, a universe cannot be born with them. Therefore, a universe should be born to contain 100 percent dense-matter objects. Besides, all universes must have an energetic atmosphere of neutrinos and ethers to make dense-matter objects unstable undergoing nuclear reactions to make matters having atomic and molecular structures and to release the huge amount of mass and energy stored in them. Dense-matter objects therefore are the starting materials for making galaxies and universes like ours not hydrogen and helium. It also means that all kinds of stars should be made of 100 percent of dense-matter objects to begin with.

Formation of a Galaxy and a Universe

Because collision interactions produce all phenomena, the formation of a galaxy or universe should also be the results of collision interactions. Findings suggest that most stars in a galaxy appear to have about the same age, indicating that a galaxy may be born simultaneously. It means that, at the beginning, there were already billions of pieces of dense-matter objects to begin their nuclear reactions simultaneously on their surfaces to make hydrogen and helium. The nature way to produce a galaxy therefore is likely a violent collision between two huge dense-matter objects, breaking both of them into billions of pieces and simultaneously producing a dense cloud containing the dust of dense-matter objects, neutrinos, ethers, electrons, protons, and huge amounts of energy. Under the bombardment of the energetic cloud particles produced by the violent collision, the dust particles and broken pieces of the dense-matter objects become "radioactive," starting nuclear reactions on their surfaces to quickly consume their dust and small-size particles to produce cosmic hydrogen and helium clouds and an atmosphere of energetic neutrinos and ethers.

Meanwhile, large broken pieces of dense-matter objects become baby stars by constantly undergoing nuclear reactions on their surfaces. The simultaneous formation of billions of baby stars producing an atmosphere of energetic neutrinos and ethers marks the birth of a baby galaxy. The newly born baby stars are essentially invisible and small because they are 100 percent dense-matter objects, which, without having atomic and molecular structures, do not emit other lights besides x-ray. A baby star is an antigravity star because it constantly shoots out energetic particles pushing everything away. Therefore, at the beginning, a baby galaxy may uniformly expand to spread their baby stars apart by their antigravity force pushing one another away.

Outside the universe, dense-matter objects still have weak gravitational force affecting one another; therefore, some may exist in group-orbiting systems. When such two groups of dense-matter objects run into each other, chances are, many collisions will occur over a long period of time with each collision forming a galaxy, and a universe may form this way over a rather long period of time.

Galaxies inside a universe therefore have different ages. Some are young; some are old. Some may have been dead before a new galaxy is born. A universe therefore does not have a meaningful age, but its galaxies do. A galaxy dies when its dense-matter objects are about to be used up. Besides, one or a group of dense-matter objects may run into a universe. A collision of a dense-matter object with a star would produce unbelievably huge amounts of energy and a fantastic view of explosion of the star.

Baby stars are 100 percent dense-matter objects. Inside a universe under constant bombardment of energetic atmospheric neutrinos and ethers, dense-matter objects become radioactive and constantly undergoing nuclear reactions to make hydrogen and helium to become stars. Findings that all stars undergo constant rate of nuclear reactions suggest that these nuclear reactions only occur at the surface of dense-matter objects, proving that even the energetic atmospheric neutrinos and ethers of the universe cannot penetrate deep into dense-matter objects due to their amazingly dense mass.

Formation of Regular Bright Stars

The new mechanical-physics-based cosmology has a very different model of universe from that of modern cosmology: an open, not energy and matter conserved, not expanding universe v. a closed, energy and matter conserved, expanding universe. Besides, the new cosmology has identified a new antigravity matter called dense-matter object by the author, which the nature produces and stores outside the universe at temperatures of zero degrees Kelvin and below. According to the new cosmology nature uses it to make stars, thus, the star models of the new cosmology are also very different from those of modern cosmology.

For example, a young regular bright of the new cosmology has a very large bright hollow outer-layer containing mainly hydrogen and helium and a dense-matter object at its center. The mass content of the outer-layer continues to increase with aging at the expense of the dense-matter object in the center constantly undergoing nuclear reactions. Unlike the model of modern cosmology to fuse hydrogen into helium the nuclear fusion reactions are those combining protons and electrons to form hydrogen and helium.

Having Artificially Low Mass Density

A baby star of a newly formed galaxy is born to have 100 percent dense-matter object, which is tiny in comparison with bright stars, but it is the densest stars having the strongest antigravity and is essentially invisible due to both that it does not emit visible lights and it is very small. Although baby stars constantly emit x-ray, they are usually too small and too far away to be detectable.

A baby star's surface constantly undergoes nuclear reactions, making both hydrogen and helium and throwing them out to a great distance, where they gradually accumulate, forming a large, hot, ball-shaped, hollow outer-layer, which gets brighter with increasing concentration and thickness to gradually turn the invisible baby star into a regular bright star. A regular bright star is therefore made of a dense-matter object at center surrounded by a very large, hollow, bright, ball-shaped outer-layer made of hydrogen and helium. Its outer-layer gets denser and thicker upon aging at the expense of the dense-matter object at its center. A regular bright star therefore looks very large, having an artificially low mass density.

Turning from Antigravity Stars to Gravity Stars

Constantly heated by the nuclear energy of its dense-matter object at center, the outer-layer of a regular star is hot and bright. Because it is made of matters having atomic and molecular structures, it has gravity by reducing the kinetic energy of neutrinos and ethers passing through it. A regular star is therefore made of an antigravity dense-matter object at center and an outer-layer made of gravity matter. Upon aging, the mass of its outer-layer continues to increase at the expense of its dense-matter object. And as a result, it gradually turns from an antigravity star to gravity star. A mature regular bright star like our sun has already turned to a gravity star. Contrary to the general belief that a regular star neither has a constant gravity nor the strength of its gravity is proportional to its mass content.

Making Elements Heavier than Helium

Because a dense-matter object constantly throws out the hydrogen and the helium it makes, it does not make elements heavier than helium, and it should be the role of the outer-layer of a regular bright star to do so. As the outer-layer gets thicker, it gets hotter, particularly its inner surface. When the temperature on its inner surface reaches the critical point, it begins nuclear reactions to make elements heavier than helium, likely beginning sometime during maturing age. Once started, the nuclear reactions occur on a very large area to make heavier elements; thus, it greatly accelerates the aging process of bright regular stars.

Building Up Inner Gas Pressure

The hydrogen, helium, and their ions constantly produced by the dense-matter object of a young regular bright star are readily soluble in its outer-layer. However, after its inner surface starts nuclear reactions to make elements heavier than helium, it begins to be coated by those heavier elements it continues to make. As the coating of heavy elements gets thicker, it gets harder and harder for the particles continuously produced by its dense-matter object to penetrate through and to dissolve into. Therefore, as a mature regular bright star ages its inner gas pressure continues to build up, and it will get to a point that it can only periodically release its gas with bursts similar to the release of solar wind by our sun.

As the inner pressure continues to build up, a regular bright star will eventually explode or violently expand to become a red giant to release its inner gas pressure as astronomers have discovered.

Dense Stars

Black Holes: Very Large Dense-Matter Objects

Very large pieces of dense-matter objects may be present particularly at and near the center locations of galaxies. Their antigravity may be so strong and the environment around them may be so hot that the hydrogen and helium gases they constantly produce continue to disperse into cosmos instead of forming giant regular stars. Because their antigravity particles destroy and disperse lights around them, the space around them looks black; therefore, they have been called "black holes." Black holes are therefore very large dense-matter objects having very strong antigravity, not gravity. This explanation, of course, counters the current concept of black holes having infinitely strong gravity from which even light cannot escape. Having irregular shape and constantly undergoing nuclear reactions on their surface some dense-matter objects are expected to fast-rotate and to do fast circular motion as well. These fast rotating dense-matter objects produce strong spiral wind storms of neutrinos and ethers containing nuclear-reaction particles surrounding them blowing their hydrogen, helium, and ion particles out to a great distance. Therefore, some black holes are made of fast-rotating large dense-matter objects throwing their hydrogen and helium out to a great distance to form giant bright rings looking like donut instead of forming gigantic regular bright stars such as the image obtained by Hubble Space Telescope in the center of galaxy NGC4261NASA. Constantly pushed by the spiral windstorm of neutrinos and ethers containing nuclear reaction particles, these bright, donut-shaped rings rotate.

Neutron Stars: Dense-Matter Objects

The new cosmology supports the teaching that large regular bright stars undergo supernovae explosion to form "neutron stars," which, however, should be the exposed dense-matter objects becoming newborn baby stars, not dead, dense stars. A supernovae explosion effectively removes the outer layer of a large regular star to expose its dense-matter object becoming an essentially invisible reborn baby star known as a neutron star. A neutron star therefore is a dense-matter object; thus, it is an antigravity star. And like black holes, they are the densest stars. They are by no means dead stars because their surface constantly undergoes nuclear reactions producing hydrogen and helium, throwing them out to a great distance gradually forming a bright outer-layer, becoming regular bright stars again.

Many neutron stars have been found fast-rotating and been called pulsars. Having both irregular shape and nuclear reactions on their surface, some are expected to fast-rotate and even simultaneously do fast circular motion as well.

White and Brown Dwarfs

Medium-sized regular bright stars such as our sun may end their life by violently expanding to become red giants and then exploding to shed a large portion of their outer-layer, leaving behind a new outer layer having greatly reduced size, which, according to astronomical findings, contains mostly carbon and oxygen (on their surface). Findings show that white dwarfs and even brown dwarfs are much smaller stars than regular bright stars but having mass densities much larger than them and the matters having atomic and molecular structures, proving they have a dense-matter object at their centers.

Findings also show that the outer-layer of white dwarf stars are much hotter than those of regular bright stars apparently due to the closeness of their outer-layers to their dense-matter objects and that their outer-layers contain elements heavier than helium. Being extremely hot, the inner portions of the outer-layer of white dwarf stars should undergo nuclear reactions to produce even heavier elements than oxygen, making their outer-layers more and more effective in blocking the nuclear-reaction gas particles constantly produced by their dense-matter objects to penetrate through and to dissolve into.

As a result, the inner gas pressure continues to build up, and white dwarfs stars are expected to violently explode again. They are known to undergo type 1a supernova explosion, the most powerful of all the supernovae. Currently, scientists believe that white dwarfs are dead stars having no nuclear fuel left. Actually, they are very active stars constantly undergoing nuclear reactions both on the surface of their dense-matter objects at its center and the inner portion of its outer-layer, finally resulting in a violent explosion to end their lives.

Brown dwarf stars are made of hydrogen and helium, having a small dense-matter object at their centers, which are significantly smaller than those inside regular bright stars and even white dwarfs. When their dense-matter objects become too small to be able to undergo nuclear reactions, they cool off to become brown dwarf stars. However, their dense-matter objects continue to be radioactive, constantly producing large amounts of energy. The antigravity of their dense-matter objects is likely too weak to separate themselves from their outer layers by a hollow space like regular stars. Once their dense-matter objects are used up, they cool down to become planets, having a low mass density like Jupiter and Saturn.

The Movement of Galaxies

Because all stars continuously produce neutrinos and ethers, each galaxy should produce a neutrino-and-ether wind constantly blowing out of the galaxy in all directions. There are many galaxies, and so are their winds. When their winds encounter one another, many spiral or rotational winds of neutrino and ethers, similar to hurricanes on the earth, should be routinely present in the universe.

The winds of neutrinos and ethers should play a major role in affecting the movements of stars and galaxies. This is evident from the findings that there are many spiral galaxies, which are rotating. Astronomers have found that the movement of stars in a spiral galaxy does not follow the rules of the movement of the planets in the solar system, which is believed to be a gravitational force-controlled system.

For example, in a spiral galaxy, the stars closer to the center of the galaxy move at about the same velocity as those farther away from the center. These findings are strong scientific evidence that the movement of stars in a galaxy is mainly controlled by the wind of neutrinos and ethers not by gravitational force, as physics community believes. Therefore, there is no need to postulate the presence of dark matters and dark energy.

Hurricanes are formed with spiral winds, and their shapes are amazingly similar to those of spiral galaxies. The wind speed inside a hurricane is about the same; therefore, its speed can specify the strength of a hurricane. Similarly, the circulating speeds of the stars are about the same inside a spiral galaxy.

Solar and Planetary Motions

Both neutrinos and ethers are able to penetrate through matters and to effectively collide interacting with their subatomic particles to produce all universal phenomena such as gravity. The atmospheric neutrinos and ethers therefore should have drag effect on all moving matters, slowing them down. The drag effect should play an important role on why all satellites and space stations only stay in orbit for a few years and eventually fall back to Earth and slow down spaceships.

One of the biggest secrets of the nature is that planets in the solar system have steady-state orbits around the sun without either slowing down or falling into the sun. It gives the physics community the strongest reason to believe and teach that neutrinos do not collide interacting with matters and, therefore, they have no drag effect on matters in motion. However, the solar system is just too orderly for us to believe that it occurs randomly, and some people even believe that highly intelligent beings design it. Scientists cannot explain why all planets orbit the sun in the same direction and even essentially on the same plane. They also cannot explain the stars' movements of the Milky Way galaxy and in other spiral galaxies.

Logically, to have steady-state orbit, planets need a constant supply of energy to overcome both the drag effect of the atmospheric neutrinos and ethers and the sun's gravitational force. The only scientific possibility is that the sun somehow constantly provides the energy or force needed by all planets and their moons to maintain their steady-state orbits on a plane. The new cosmology discovered by the author now offers such a scientific explanation.

Both neutron stars and black holes are dense-matter objects, and many have been found fast-rotating and are expected to do circular motions, having very strong magnetic force fields around them. Therefore, many dense-matter objects inside regular bright stars such as our sun may also be fast-rotating and simultaneously do circulating as well. Because the dense-matter object inside a regular bright star has a large space separating it from its outer-layer, its fast rotation and circular motion do not significantly affect the shape of its outer-layer, which we see as a large, spherical bright star.

Because the outer-layer of a mature regular star like our sun has strong gravity, it already has overcome the antigravity of its dense-matter object at center to become a gravity star. However, the antigravity effect of the dense-matter objects inside those regular bright stars such as our sun have not been completely overcome by the gravity of its outer-layer in all directions due to their fast rotation. A fast-rotating dense-matter object may also do fast circular motions, creating a spiral neutrino-and-ether windstorm around it, containing the gas and their ion particles the dense-matter object constantly produces by its nuclear reactions, including protons, electrons, hydrogen, helium, and their ions. This spiral windstorm pushes the outer layer of the star to rotate.

The outer-layer of the star blocks all storm particles, but allows the spiral wind of neutrinos and ethers to penetrate through with reduced strength but still able to provide a a directional spiral wind of neutrinos and ether around the star such as our sun. This spiral wind produces both an antigravity force and an anti-drag force to support those planets and their moons orbiting in the direction of the wind to obtain steady-state orbits.

If the sun catches a planet in the general direction its spiral neutrino-and-ether wind blows, it constantly offers the planet a push to support its steady-state orbiting motion. Such a spiral wind should have much stronger pushing force near the star and gets weaker quickly outward. Planets closer to the sun, therefore, should orbit faster and need stronger push to overcome both the gravitational force of the star and the drag effect of the atmospheric neutrinos and ethers to stay in orbit than the planets farther away.

Of course, all planets must orbit in the same direction of the spiral neutrino-and-ether wind to survive. Contrary to general belief that some planets were formed during the formation of stars, the new cosmology teaches that all planets of stars should be caught by them one at a time after they have turned into a gravity star. Also, all the planets of the sun should not orbit on the same plane to begin with. It must have taken a long time for all of them to gradually adjust their own orbiting direction to the direction the dense-matter object rotates and circulates or the direction of the star rotates, resulting in reach a final equilibrium having all planets almost orbiting on the same plane.

Finding no planet orbiting the sun in clockwise direction indicates that those planets have not survived. Statistically, our sun may have captured about ten planets and even many more moons circulating in the direction against the spiral wind and having plunged into the star. Once the sun's gravity catches a new planet, it will take a long time for all its planets to reach a new equilibrium of orbits.

It is also reasonable to conclude that the same spiral neutrino and ether wind keeps the moons of planets in orbits as well. Most moons also orbit in the same counterclockwise direction around their planets with a few exceptions, which are the outer moons of Jupiter (Anake, Carme, Pasiphae, and Sinope), of Saturn (Phoebe), and Neptune (Triton). However, they are all circulating the sun counterclockwise.

The force of the spiral neutrino-and-ether wind of the sun should have two components: the anti-drag force in the planet orbiting direction and the anti-gravity force component pushing planets away from sun. However, because all planets are in a steady-state orbit with all acting forces at equilibrium, we do not feel any force acting on Earth.

A satellite or space station orbiting in the direction of the Earth and moon around the sun should have longer lifetime than those orbiting in the opposite directions. In theory, those having maximized its utilization of the spiral neutrino wind of the sun should be able to stay in orbit essentially forever, like what planets and their moons do.

The same spiral neutrino-and-ether wind of the sun should also provide the planets with rotational energy. The storm particles, such as protons, electrons, hydrogen, helium, their ions of the spiral wind of neutrinos, and ethers, provide stronger force than the wind itself to push the outer-layer of the sun to rotate. The rotation of our sun therefore proves that its dense-matter object undergoes nuclear reactions and is fast-rotating.

Capable of offering logical explanation of the motions of the sun, its planets, and most of their moons further proves that the mechanical physics-based new cosmology is scientifically correct.

Discussion

Natural Science: The Only Correct or Real Science

Mechanical physics has long been applicable to everything—the foundational science of all fields of natural science; however, the physics community has barred its use in interpreting universal phenomena, cosmology, particle physics, and related scientific topics in order to further its development and the teaching of mathematical physics or mainstream-of-thought physics. Since Einstein's scientific revolution, the physics community has claimed that mechanical physics is incapable of interpreting relativistic phenomena and later claimed it is also unable to interpret quantum phenomena. It therefore has led the scientific community to develop mainstream-of-thought physics to replace mechanical physics to interpret these phenomena and insist that mechanical physics is not the physics of everything or the only correct physics.

If mainstream-of-thought physics were the correct science, the mechanical physics-based interpretations of the above-mentioned scientific topics already interpreted by the mainstream-of-thought physics should never be found, but the author has discovered them. Due to their incompatibility, the scientific interpretations of all scientific topics by the two physics should be entirely different, but only one of the two physics can be scientifically correct. If the interpretations of the above scientific topics by mainstream-of-thought physics were proven correct as the physics community claimed, the entire mechanical physics, including the entire natural science, should be wrong and nonexistent, but this is apparently impossible.

Now, the breakthrough discoveries of the author have found all the mechanical physics-based interpretations of the above-mentioned scientific topics already interpreted by the mainstream-of-thought physics, which prove again that mechanical physics is the physics of everything and natural science is the only correct or real science. It is true, as the physics community claimed, that mechanical physics has no scientific interpretation for relativistic phenomena, the quantum phenomenon of universal forces, and all the elementary particles predicted by standard model particle physics. They however are not needed by the mechanical physics to interpret the above scientific topics, thus, having been proven not needed, nonexistent, and obsolete.

Mainstream-of-Thought Physics: Not Natural Science!

The natural science discovered by Copernicus and Galileo teaches using experimental findings to discover natural laws. Both matters and the phenomena they produce are experimentally detectable, quantifiable, and can be studied and discovered experimentally because they undergo collision interactions, resulting in reflecting lights, producing energies, and/or forces detectable experimentally. Because matters produce all phenomena such as light, energies, and forces detectable experimentally, anything detectable experimentally must also be matters, which have unique features: having mass and relativity motions, taking up space, and undergoing collision interactions.

The above features of matters lead to the specific definition of natural science: natural science study matters and the phenomena they produce experimentally. This definition has excluded anything or any force or energy without matter or mass to be natural science.

Our physics community claims that mainstream-of-thought physics requires experimental proof of its mathematically predicted universal phenomena and elementary particles; therefore, it is natural science, but it cannot explain why mathematical derivation can make scientific discovery, which therefore is a religious belief. Physics community also claims that electromagnetic wave has been proven to be light, but without having mass, how electromagnetic wave can be experimentally detectable or proven to be light.

Developed with postulations contradictory with natural science, all the scientific interpretations of mainstream-of-thought physics should be contradictory with natural science or experimental findings. Therefore, it cannot be natural science and proven correct experimentally.

Therefore, once the definition of the natural science is strictly applied, Einstein's theory of relativity fails the test of being natural science because its mathematical origins such as its postulated force fields, electromagnetic wave, spacetime, curved spacetime, and so forth have no mass and therefore should not exist and be detectable experimentally. This means that light cannot be electromagnetic wave and gravity cannot be curved spacetime because both light and gravity are experimentally detectable; thus, they should have mass.

Because light cannot be electromagnetic wave, the predicted relativistic phenomena based on the postulations that electromagnetic wave were light cannot be true to exist. It means that there is no scientific reason to believe that mathematical derivation can predict or discover universal phenomena and elementary particles. Therefore, the entire mainstream-of-thought physics using mathematical derivation to make scientific discovery is not the natural science making scientific discovery experimentally.

Mainstream-of-Thought Physics: Teaching a Fantasy Universe

Author's breakthrough scientific discovery has shown that mechanical physics can interpret all the scientific topics already interpreted by the mainstream-of-thought physics disproving all the claims of the physics community that mechanical physics is incapable of doing so. It therefore has overcome the repeated claims of the physics community to prove both that mechanical physics is the physics of everything and natural science is the only correct or real science.

Findings that all universal forces are explainable by act-on-contact collision interactions of the atmospheric neutrinos and ethers with matters to produce act-on-contact pushing forces have finally proven that act-from-distance attraction universal forces do not exist and are obsolete. Therefore, Faraday postulated matter-free force-fields do not exist. And it further means that their force mediating particles, spacetime, curved spacetime, the universe represented by Einstein's field equations, and so forth do not exist. Therefore, the physics of attraction forces acting from distance or the mainstream-of-thought physics has been proven wrong teaching a fantasy universe doesn't really exist.

Both general theory of relativity and big bang theory are the foundation science of modern cosmology but general theory of relativity has been disproved and its spacetime has been proven not to exist. Big bang theory is based on the scientific interpretation of the phenomenon of universal redshift of starlight that the nature gave birth to the universe with a big bang or explosion of a "singularity" about fifteen billion years ago and is uniformly expending ever since. The author however has shown that his new ether theory not only can interpret the phenomenon of universal redshift of starlight but also many other light-related universal phenomena consistently. And therefore interpretation of the redshift of starlight by the new ether theory is supported by overwhelmingly more findings than the big bang theory, thus, proving big bang theory scientifically wrong. Therefore, the entire modern cosmology has been proven wrong scientifically. The discovery of a new mechanical physics-based cosmology by the author further proves that modern cosmology is wrong.

Over the past century, both some centuries-old scientific misconceptions and Einstein's theories of relativity have misled the physics community to lead the world to develop a religious pseudo-physics known as the mainstream-of-thought physics teaching a four-dimensional fantasy universe based on beliefs that mathematical derivation makes scientific discovery. Although it is now proven that everything it teaches is wrong scientifically and a fantasy universe does not really exist, it has been accepted and recognized as the correct science or natural science by the entire scientific community and all nations worldwide.

Apparently, the highest trust in scientific communities by the general public has been exploited by the physics community, which has been misled and/or corrupted by Einstein's theories of relativity and old scientific misconceptions. Apparently, the physics community has made a big scientific mistake a century ago to support Einstein's theories of relativity and even to lead his scientific revolution even when these theories are based on postulations contradictory with natural science, the repeatedly proven correct science for centuries. It has been the repeatedly false claims of the highly trusted physics community that the mainstream-of-thought physics has been proven correct experimentally convinced the entire scientific community to accept and support it as the proven correct science.

The author calls the mainstream-of-thought physics religious pseudo-physics since it is based on beliefs that mathematical derivation makes scientific discovery, is not logically understandable, and nobody really understand it. Also, the scientific communities of those scientific fields besides physics and cosmology do not really support the mainstream-of-thought physics. This is apparent since their sciences are still parts of natural science making scientific discovery experimentally. Even in the scientific field of physics mechanical physics is still the daily applied physics and the foundation science of all other fields of natural science.

Therefore, only a very small portion of the scientific communities in physics, cosmology, and astronomy has been misled and/or corrupted to support the development and the teaching of religious pseudo-physics in the past century. The majority of the scientific community continues to develop and teach the mechanical-physics based natural science. Intimidated by difficult mathematics and not been able to understand logically have made most scientists silent allowing the corrupted physicists to develop, teach, and fraudulently misuse scientific community, R&D and science education funding to develop and teach religious pseudo-physics in the past century.

The author and the general public trust that the scientific community and the scientific agencies of governments worldwide are obligated to develop, support, and teach the only correct or real science. He therefore advocates both scientific community and governments worldwide to seriously face the author's allegations that the entire mainstream-of-thought physics is scientifically wrong teaching a religious fantasy world doesn't really exist.

Over the past century, there have been many scientists, even scientific organization, and journals dedicating to disprove the mainstream-of-thought physics. Many alternative theories have been proposed to modify or replace some of its theories. Also, due to that mathematical theories are non-specific to any scientific interpretation, many mathematical theories have been proposed to compete with the theories of the mainstream-of-thought physics.

However, surprisingly, the author is unique and alone to insist that the classical or mechanical physics should be the only correct or real science. He is also unique to offer the expanded mechanical-physics based natural science to not only disprove but also to replace the entire mainstream-of-thought physics. Besides, his breakthrough discovery opens an advanced broad-scope scientific understanding of the natural and hopefully will lead to scientific breakthroughs and advancements to brighten our future.

www.ingramcontent.com/pod-product-compliance
Lightning Source LLC
Chambersburg PA
CBHW022101170526
45157CB00004B/1435

* 9 7 8 1 4 8 3 4 1 4 9 6 6 *